埃 萨 · 皮 罗 宁

Esa Piironen

图书在版编目（CIP）数据

埃萨·皮罗宁／方海主编．—北京：中国建筑工业出版社，2004
（欧洲当代著名建筑师作品精选）
ISBN 7-112-05131-2

I.埃… II.方… III.建筑设计-作品精选-欧洲-现代 IV.TU206

中国版本图书馆CIP数据核字（2003）第107815号

Images and texts, Esa Piironen

主编：方海
Chief-editor: Fang Hai

翻译：单皓
Translator: Shan Hao

责任编辑：黄居正　徐纺　曹扬
Editors: Huang Juzheng, Xu Fang, Cao Yang

责任设计：刘向阳
责任校对：赵明霞

欧洲当代著名建筑师作品精选
埃萨·皮罗宁
方海主编　单皓译
*
中国建筑工业出版社出版、发行（北京西郊百万庄）
新华书店经销
伊诺丽杰设计室制版
精美彩色印刷有限公司印刷
*
开本：787×1092毫米　1/12　印张：19⅔
2004年5月第一版　2004年5月第一次印刷
印数：1—2,000册　定价：159.00元
ISBN 7-112-05131-2
TU·4555（10745）

本社网址：http://www.china-abp.com.cn
网上书店：http://www.china-building.com.cn

方海　主编

欧洲当代著名建筑师作品精选

埃萨·皮罗宁

Esa Piironen

中国建筑工业出版社

CHINA ARCHITECTURE & BUILDING PRESS

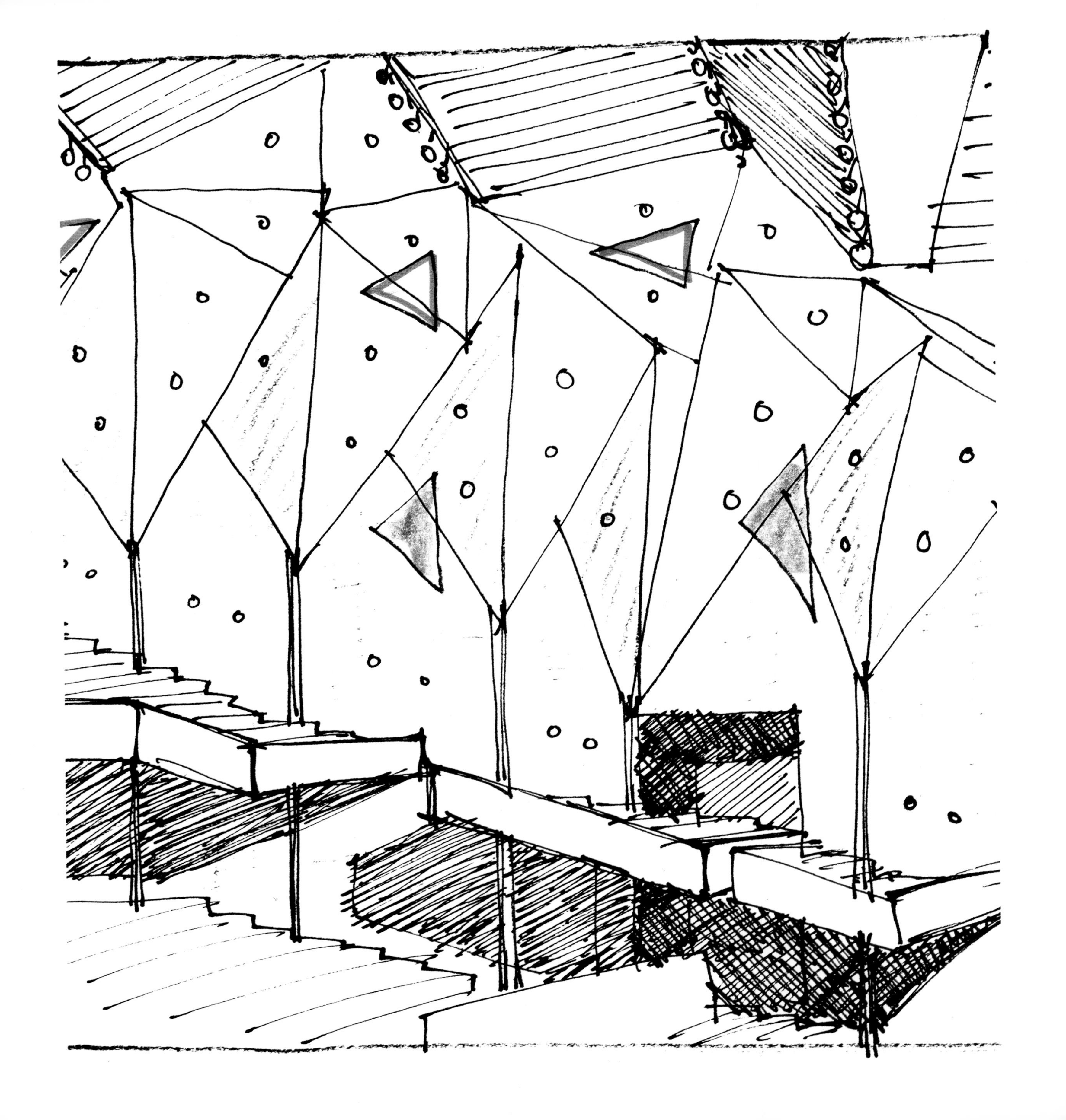

目　录　　contents

建筑师埃萨·皮罗宁访谈

采访人：方海

1

你能谈谈你家庭的情况吗？在你成长为一名建筑师的过程中，是否曾受到过你父母的影响？

我父亲出生于卡累利阿，那里曾经是芬兰的一部分，战后划归苏联。他在航行于世界各地的商船上担任轮机长，几乎一辈子都在海上漂泊。有时候我也有机会跟着他跑跑。我母亲出生于芬兰中部，她的亲人都是农民。

我父母并未影响我成为一名建筑师，幸运的是，当我15岁时曾作为一个夏日学徒在佩卡·皮特凯宁建筑师事务所实习，一开始仅仅是准备咖啡、传递文件等，后来则被允许描图和制作模型。

2

你什么时候设计了你的第一座建筑？它是什么样子的？

在建筑事务所工作的时候，我在图尔库群岛为父母设计了一座度夏小屋。它的结构很简单，我父亲把它盖了起来。眼看着自己的设计被盖起来是件很有趣的事情。在那一刻我确信自己想成为一名建筑师。

3

你所进的建筑学院是什么样的？你主要受到的是古典建筑教育还是现代建筑教育？你喜欢教你的老师吗？你认为哪一位老师对你最重要？能不

1

Interview

By Fang Hai

1

Can you tell me something about your family? Did your parents influence you to become an architect?

My father comes from Karelia, which became part of the Soviet Union after the Second World War. He worked almost all his life as a chief engineer on merchant ships sailing all over the world. Sometimes I had an opportunity to go with him. My mother comes from central Finland from a family of farmers.

My parents did not affect my choice to become an architect. I was lucky though to work at the Architectural Office of Pekka Pitkänen in Turku since the age of 15 as a summer trainee. First I just did some chores like making coffee and delivering copies, but later I was allowed to do some drafting and make models.

2

When did you design your first building?

While working at the architectural office I designed a summer cottage of a simple structure for my parents in Turku Archipelago. It was very interesting to see your own design being built. At that time I was certain I wanted to become an architect.

3

What was the architectural school like when you entered ? Did you have mainly a classical architectural training or modern training? Do you like the teachers who taught you Who do you think was the most important teacher for you? Can you tell me who else had great influence on your development as an architect?

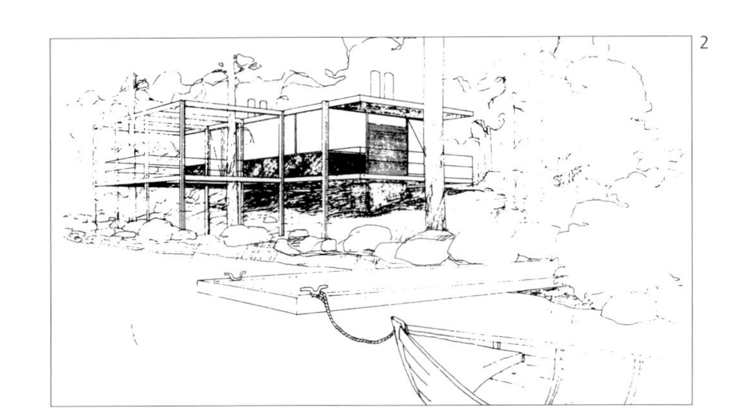

2

1 从赫尔辛基西那塔广场所绘的图，1962 年
2 托贝格别墅，福贝，方案，1967 年
3 Lounaisrannikko 台式住宅，埃斯波，竞赛入围，1968 年
4 模度住宅，NCSU，方案，1972 年

1 Drawing from Helsinki Senate Square 1962
2 Villa Troberg, Förby, project 1967
3 Lounaisrannikko Terraced Houses, Espoo, competition entry 1968
4 Modular House, NCSU, project 1972

能告诉我还有谁在你成长为建筑师的过程中起了很大影响？

因为 1959～1961 年的夏天我在图尔库的建筑事务所干过一段时间，所以很容易就取得了赫尔辛基建筑学院的入学资格。我们连着考了五个礼拜。我们既要画出指派的任务，又要设计出这些任务，还得考数学。

那时候赫尔辛基技术大学是全国惟一的一家建筑院校。这里有一套使用多年的培养建筑师的课程。你在学习建筑学的同时还得学城市规划以及一切与建筑有关的知识。这里所学大多是现代课程，气氛宽松而热烈。对我来说，最重要的老师无疑是阿尔诺·鲁苏武奥里。他担任低年级的建筑教授。最初的两年中我们设计了所有类型的小房子。他是位迷人的讲授者，介绍我们认识了阿尔瓦·阿尔托、勒·柯布西耶以及密斯·凡·德·罗的建筑。他们在我成长为一名建筑师的过程中产生了巨大影响。随后还有路易斯·康、巴克敏斯特·富勒和弗兰克·劳埃德·赖特以及其他一些人。鲁苏武奥里最喜欢一种芬兰式的极少主义风格。

技术大学还有一位出色的建筑学教授——奥利斯·布卢姆斯泰特。因为我在低年级，他并没有实际教过我什么课，但有时候我喜欢跑去听他的建筑哲学课。他们给我的建筑思索带来一些相当新颖的想法。

As I had been working at the architectural office in Turku in summers 1959–61 it was quite easy to be qualified for the architectural school in Helsinki in 1962. We had tests that lasted for five weeks. We had to do drawing and design assignments as well as tests in mathematics.

The Technical University in Helsinki was the only architectural school in the country at that time. It had the same curriculum to train architects for years. You had to study city planning as well as architecture and all aspects related to building. In a way it was quite a modern curriculum and the atmosphere was very enthusiastic. The most important teacher for me was no doubt Professor Aarno Ruusuvuori. He was a professor of architecture for beginners and the first two years we designed small houses of all kinds. He was a fascinating lecturer and introduced us to the architecture of Alvar Aalto, Le Corbusier as well as Mies van der Rohe. They had a great influence on my development as an architect. Later came Louis Kahn, Buckminster Fuller, Frank Lloyd Wright and others. A kind of Finnish minimalism was Ruusuvuori's great interest.

At the Technical University another remarkable professor of architecture was Aulis Blomstedt. I was still at a lower level and he did not actually teach me, but I used to go and listen to his philosophical lectures on architecture. They gave remarkable new ideas to my architectural thinking.

4

新的构思来自何处？能不能告诉我你的设计过程，也就是你创造作品的方式？你是画许多草图呢，还是在你动笔之前先想好一个构思？或者你主要靠做模型？你经常使用特别的模数吗？

4
Where do new ideas come from? Can you tell me about your design process: the way you produce your works? Do you sketch a lot, or do you think right through an idea before you begin to draw? Or do you mainly make models? Do you usually use a particular module?

3

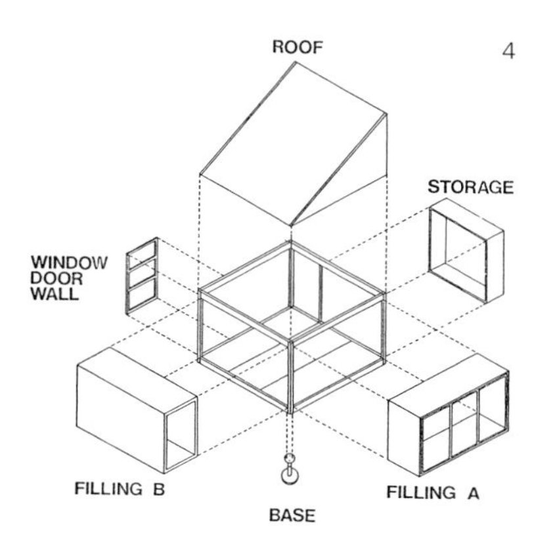

4

新的构思来源于不同的层面。不同季节的芬兰自然景观是影响力最大的一个源泉。旅行的时候我喜欢观察老房子和新房子。另外当然会仔细研读建筑论著和杂志。几乎所有的构思早就已经存在某处了，但主要靠你如何用一种新的方式转换和发展它。

根据我在何时设计何种建筑的具体情况，我采用不同的方法。我们通常会画很多图，有些甚至是错的或者画不出来的。有时候基本构思在你的第一次草图中就已经出现了。你需要一遍又一遍地重画。我们一般一开始就做模型，有些工程在设计中要做 1：20 或 1：50 比例的模型。

我们在设计中并不使用特别的模数。有时模数就是1mm甚至更小的单位。在图形设计中模数称作“pica”，表示为 0，000mm。

5

你如何成立了自己的事务所？你能不能给我谈点关于你事务所的构成情况？在你的事务所中职责是如何分派的？

1966 年我和米科・普尔基宁组建了自己的第一家事务所，那时我们还是学生。我们都在图尔库的佩卡・皮特凯宁事务所工作。我们从客户、亲戚和朋友那里得到了一些自己的工程。它们大部分是小房子和度夏小屋，对后来的工作来说是很好的锻炼。1969 年我搬到赫尔辛基，完成了学业，事务所一直维持到 1971 年。

1970 年我和奥拉・莱霍、埃斯科・米耶蒂宁、尤哈尼・帕拉斯马一起成立了一家新的事务所。那家事务所的名叫“Suunnittelutoimisto G4”。

New ideas come from different sources of course. Finnish nature with its different seasons is a great source of inspiration. When travelling abroad I like to see both old and new buildings alike. And of course I study architectural literature and magazines. Almost all ideas are already there, somewhere, but it depends how you transform and develop them in a new way.

I use different types of approach depending on the situation when and what I design. I usually draw quite a lot, even impossible sketches. Sometimes the basic idea is already in your first drawing. You just have to redraw, again and again. We usually make models right from the beginning, and some projects were designed with models in a scale 1:20 and 1:50.

We do not use any particular module in our design. Sometimes the module is one millimeter or even less like in graphic design.

5

How did you set up your office? Could you tell me a little about the structure of your office? How do you delegate responsibility in your office?

I had my first office with Mikko Pulkkinen in 1966 when we were still students. We were both working at Pekka Pitkänen's office in Turku. We had some projects of our own mainly from clients, relatives and friends. They were mainly small houses and summer cottages. That was very good practice for future works. Then I moved to Helsinki in 1969 to complete my diploma work and the office operated until 1971.

In 1970 we set up a new office with Ola Laiho, Esko Miettinen and Juhani Pallasmaa. That office was called Suunnittelutoimisto G4. We started to design a new sign system for Helsinki Metro; first a research project, then a design project and finally the

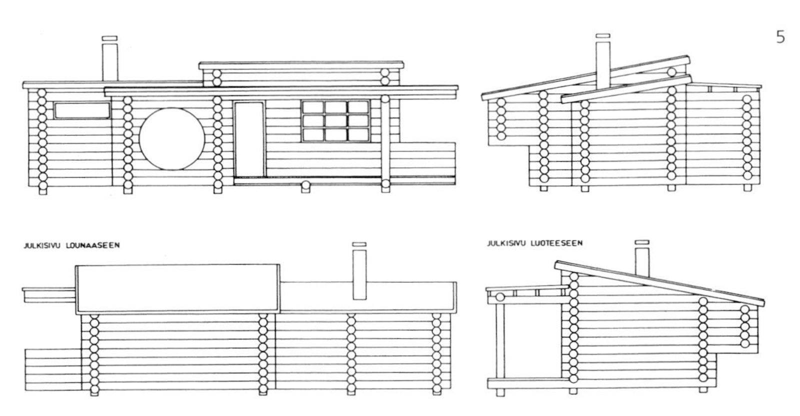

5

6

5 桑那别墅，Pakinainen方案，1975年

6 坦泊图书馆，1978年竞赛入围

7 多用途村，1979年竞赛入围

8 卡伏拉文化中心，1979年竞赛入围

5 Sauna cottage, Pakinainen, project 1975
6 Tampere Library, competition entry 1978
7 Multia Main Village, competition entry 1979
8 Kouvola Cultural Center, competition entry 1979

我们开始为赫尔辛基地铁设计一套新的标志系统。这开始是一个研究项目，然后变成了一项设计项目，最后成为实施项目。我们花了12年的时间来设计它，而赫尔辛基地铁准备于1982年投入运行。我们设计过所有类型的项目：图案设计、展览、街道设施、建筑乃至城市规划。我与埃斯科·米耶蒂宁一直合作到1985年。

在此之前，我与萨卡里·阿尔泰洛一起成立了一家新的事务所。我们都在奥坦涅米技术大学教书，我的母校搬到了那里。我们开始参加公开的建筑竞赛。在事务所的早期阶段，我们取得了一些胜利。我们赢得的第一个项目是劳塔瓦拉教堂。那是我们第一个建成的重要作品。这样就带来了其他的项目，有些项目来自于竞赛。我们主要的作品坦佩雷音乐厅就是通过竞赛获得的。我们的合作一直持续到1990年，坦佩雷音乐厅落成于当年。随后我们分别创立了自己的事务所。

在我所有的事务所中都有一条准则：每个人都要尽可能多地参与设计。所有的基本想法一般都源自合股人，但是我们会尽一切可能的力量来发展方案。事务所的规模越大，合股人控制所有细节的能力就越弱，但我们总是努力去控制所有的细节。有时候我们有20名助手，有时只有3名。通常由某一位合股人负责设计工作，但所有的构思都是共同讨论的。每一个较大的委托项目都需要一名负责工程的建筑师，但所有的问题都通过一个团队来解决。有时候我们的事务所就像一个大家庭。我们习惯于尽量分清责任，但最后的责任是由我来承担的。

implementation. It took us 12 years to design it; Helsinki Metro was ready to operate in 1982. We designed all kinds of projects: graphic design, exhibitions, street furniture, buildings and even city planning. I had the office with Esko Miettinen until 1985.

Before that I set up a new office with Sakari Aartelo in 1978. We were both teaching at the Technical University in Otaniemi, where my old school had moved. We started to participate in the open competitions for architects. We had some success during the early years of the office. Our first winning entry was Rautavaara Church. It was our first major work that was realized. Then came other works, some through competitions. Our main work is Tampere Hall, that was based on a winning competition entry. Our co-operation lasted until 1990, the year when Tampere Hall was completed. Then we both set up our own practice.

In all my offices there has been a rule that everybody can participate in designing as much as possible. All the basic ideas usually came from the partners, but the development of a project is done by all the workforce available. The bigger the office the less the partners are able to control every detail. But we always tried to control them. Sometimes we had over 20 assistants, sometimes only three. One of the partners was usually responsible for the design work, but all ideas were discussed together. Every bigger commission needed a project architect, but all problems were solved in a team. Sometimes our office was like a big family. I like to delegate responsibility as much as possible, but the final responsibility is of course mine.

6

在你作为一名建筑师的工作中，在不同的要素之间是否存在着一个关于

6

In your work as an architect, is there an order of importance between the different factors – speaking of structural order,

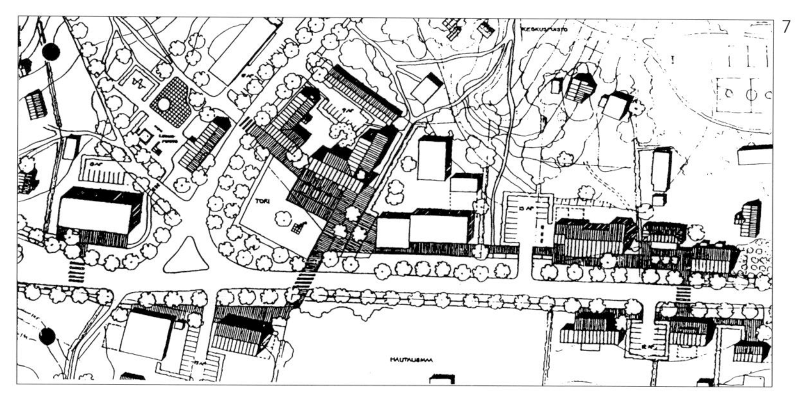

7

8

重要性的秩序？——如果谈到结构方式、空间、体量和立面，彼此是否存在先后关系，亦或设计总是绝对从功能开始？

秩序是……很重要的，正如路易斯·康曾经说过的那样。我们努力解决建筑中的复杂问题，对建筑的所有方面进行设计：场地，历史，气候，景致，技术层面，功能，计划等等。我们从哪一点出发并不重要，整体大于各部分之和。我们的目的是使建筑浑然无缺，以满足使用者的需求和我们自己确定的目标。

7

你旅行过很多地方吗？哪一次旅行对你来说最为重要？除了你前面提到的那些人外，你还和国际上的哪些人有所交往？

1951 年我作了第一次长途国际旅行，那次我和父母一起坐一艘商船去伦敦。我们在伦敦南班克逗留期间，有一个大型展览“不列颠节”，给我留下很深的印像。

当我是一名建筑学学生的时候曾经环游欧洲，参观所有的古今建筑杰作。1964 年我曾在纽约的雷诺·阿尔尼奥建筑事务所工作。在参观中我看到了不少有趣的建筑场所，并且访问了密斯·凡·德·罗和菲利普·约翰逊的事务所。1971～1972 年我在南卡罗来纳州立大学学习建筑，获得了建筑学硕士学位。之后我从北京一路旅行到新西兰克赖斯特彻奇。我女儿在美国生活，因此我对这个国家相当熟悉。

space, volume, and facade, does one of these follow another, or does the design absolutely start from function?

Order is.....very important as Louis Kahn said. We try to solve the complicated problem of architecture facing all aspects of the building to be designed: site, history, climate, views, technical aspects, function, program etc. Where we start is not so important; the whole is more than the sum of its parts. Architectural harmony is our goal, to satisfy the needs of the user and the goals we have set.

7

Have you travelled a lot? Which trip has been the most important to you? Have you any important international personal contacts besides the ones you mentioned before?

My first longer trip abroad took place in 1951, when I sailed with my parents to London on a merchant ship. There was a big exhibition during our stay at London's South Bank, Festival of Britain that impressed me.

As a student I travelled through Europe to see all the masterpieces of old and modern architecture. In summer 1964 I worked at the architectural office of Reino Aarnio in New York. I saw some architecturally interesting places during my visit and visited the offices of Mies van der Rohe and Philip Johnson among others. I studied architecture at North Carolina State University in 1971–72 and had my master of architecture degree. After that I have travelled from Peking to Christchurch, New Zealand. As my daughter lives in the USA, I still go there frequently.

9

10

9 利沙米图书馆和文化中心，1980年竞赛入围

10 波利音乐与国会中心，1981年竞赛入围

11 拉马镇市政厅，1981年竞赛入围

12 尤文帕管理和文化中心，1981年竞赛入围

9 Iisalmi Library and Cultural Center, competition entry 1980
10 Pori Music and Congress Center, competition entry 1981
11 Rauma Town Hall, competition entry 1981
12 Järvenpää Administrative and Cultural Center, competition entry 1981

8

你还接触其他艺术吗？比如绘画、雕塑、音乐或别的什么？你经常从其他艺术中汲取灵感吗？你曾和其他艺术家合作过吗？

当我还是一名建筑学学生的时候就开始从事图形设计，一直持续到今天。我设计海报、商标、书籍图案、信头、标志系统和所有的图形。在我们的建筑中，我们有机会和当代的艺术家合作。他们和建筑师一起在建筑中进行艺术创作。

劳塔瓦拉教堂中的圣坛雕塑是和艺术家阿努·马蒂拉合作设计的成果。坦佩雷音乐厅的艺术品由约尔马·豪塔拉和建筑师共同设计，但坦佩雷音乐厅中的主要艺术品"蓝色线条"由基莫·凯万托设计。同一座建筑中的其他艺术品则分别由马蒂·库亚萨洛、蒂莫·萨尔帕内瓦、伊尔玛·库卡斯耶尔维和泰穆·绍科宁承担。安基尼·卢凯拉和尤西·尼瓦分别设计了凯萨涅米（Kaisaniemi）地铁站和武奥萨里（Vuosaari）地铁站中的艺术品，后者以一次艺术设计竞赛的获奖方案为基础。目前我们正在两个艺术项目中和艺术家合作：一座是基莫·凯万托设计的雕塑，即将建于赫尔辛基火车站；另一件是汉努·西伦为赫尔辛基卡利奥社会办公大楼所设计的雕塑。

9

你如何得到用户和专业评论者对你的评价？你对关于自己作品的评论有何想法？它令人鼓舞还是令人沮丧？

在芬兰没有足够的建筑评论。我们对建筑的讨论并不多。这是一个

8

Do you have any connections with other arts? Painting, sculpture, music or others? Do you often get inspiration from other arts? Have you collaborated with artists?

As a student of architecture I started to do graphic design. That has continued until this day.

I have made posters, trade marks, graphic design of books, letterheads, signing systems and all kinds of graphic design.

In our buildings we have had an opportunity to work with our contemporary artists as colleagues.

The altar sculpture of Rautavaara Church is a result of collaboration with artist Anu Matilainen. The art work at Tampere Hall was designed together by Jorma Hautala and the architects, but the main art work at Tampere Hall is by Kimmo Kaivanto, "The Blue Line". Other art works at the same building are by Matti Kujasalo, Timo Sarpaneva, Irma Kukkasjärvi and Teemu Saukkonen. The art work of Kaisaniemi Metro Station is by Annikki Luukela and that of Vuosaari Metro Station by Jussi Niva, the latter based on a winning entry in an art competition.

9

How have you received criticism from users and professional critics? What do you think of criticism of your works? Has it been encouraging or depressing?

In Finland we do not have enough architectural criticism. We do not discuss architecture enough. That is a pity because architecture does not develop in a correct way. In architectural competitions we have a decent amount of criticism and I think our competition system in Finland is among the best in the world.

11

12

遗憾，因为建筑并未按照一条正确的道路发展。在建筑竞赛中我们对建筑有足够的评论，我想我们芬兰的竞赛体制是全世界最好的。

用户常常是最好的评论家。当然我的妻子也是。

10

你参加过许多建筑竞赛并获得了许多胜利。是否还记得在一些没有获胜的竞赛中，你曾经提出过一些未能实现的新想法？

有时候也许有吧，但一般来说如果你在一次竞赛中未能获胜，那就说明你要么是走了一条错误的道路，要么是工作还没做到家。我们以前参加过很多次音乐厅的设计竞赛，直到1984年才赢得坦佩雷音乐厅设计竞赛的胜利。

11

你是否能谈谈芬兰、欧洲以及全世界过去十年来建筑发展的趋势？近五年或十年来你的手法是否有所改变？你认为在自己的工作中那些方面比较重要？

在芬兰国内，60年代我们兴起过一阵混凝土建筑的热潮。接着70和80年代是钢结构走红。60年代以后北欧理性主义成为芬兰的主流。近几年来高技建筑发展迅猛。木构建筑也拥有自己的主顾。另外，当然生态建筑亦已流行多年。我觉得芬兰一般要比世界性的潮流要滞后

Users are often the best critics of my work and of course my wife.

10

You have participated in many architectural competitions and won a few. Do you have the impression that in some competitions, in which you were unsuccessful, you may have presented some new ideas which were not understood?

Sometimes maybe, but usually when you are unsuccessful in a competition, you have either taken a wrong approach or the work is done only half way. We had a lot of practise with music hall competitions until we won the Tampere Hall competition in 1984.

11

Can you tell about something of the trends of architecture in the past ten years in Finland, in Europe and in the world? Has your approach changed in the last five or ten years? What do you think are the most prominent features in your works?

On the national level we had a boom of concrete architecture in the sixties.

Then there was steel architecture in the seventies and eighties. Nordic rationalism has been the main trend in Finland since the sixties. High-tech architecture has developed rapidly during the last few years. Also wooden architecture has had their sponsors. And of course ecological architecture has been coming for many years. I think we in Finland are usually some years behind the

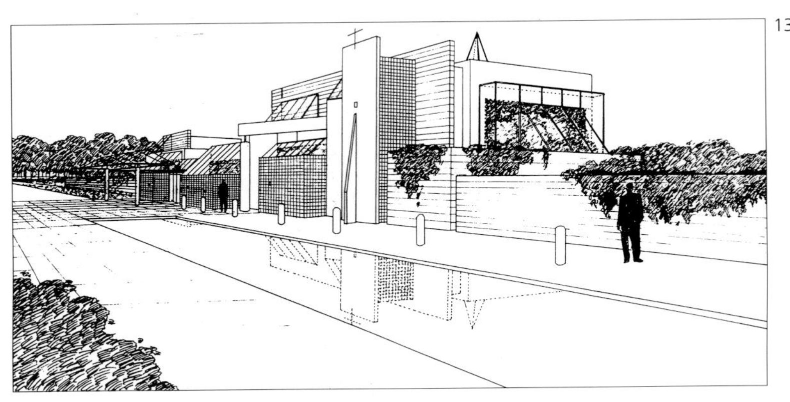

13

14

13 兰敏帕墓园和教堂，塔佩雷，1986年竞赛入围
14 兰敏帕墓园和教堂，塔佩雷，1986年竞赛入围
15 赫尔辛基大学Kumpula，化学楼，1988年竞赛入围
16 塔佩雷市政厅，大观众厅，1989年

13 Lamminpää cemetery and chapel, Tampere, competition entry 1986
14 Lamminpää cemetery and chapel, Tampere, competition entry 1986
15 Helsinki University Kumpula, Chemistry Building, competition entry 1988
16 Tampere Hall, large auditorium 1989

一段时间。

近年来我们一般采用钢和玻璃来做材料。评价自己的作品是件很困难的事情，这应该由别人去做。

12

你如何看待埃里尔·沙里宁、阿尔瓦·阿尔托和雷马·皮耶蒂莱？你能否评论一下他们的主要作品？

他们都是国内的大建筑师，同时具有很高的国际声望。埃里尔·沙里宁的赫尔辛基火车站对我来说是非常熟悉的，因为我们有幸设计修建了这座车站的站台顶棚。那座建筑至今仍是他最好的作品之一。埃里尔·沙里宁经历了多种建筑思潮，从本国的浪漫主义直到国际式。

众所周知，阿尔瓦·阿尔托是20～70年代芬兰建筑界的关键人物。他开始时设计古典风格的作品，随后在图尔库及芬兰的其他地方设计功能主义的建筑。我出生于图尔库并在那里一直工作到1969年，所以对阿尔瓦·阿尔托的作品很熟悉。阿尔托后来发展了自己的风格，创作出许多伟大的建筑作品。他的人文主义手法一直吸引着我。武奥克森尼斯卡教堂和芬兰厅是经常让我感动的两个杰作。玛丽亚别墅在一座私人住宅中把古典和现代的意象融为一体，由此成为世界上最伟大的小住宅之一。

雷马·皮耶蒂莱出生于图尔库，但他从未有机会在故乡设计过任何东西。他是芬兰建筑的一个异数，他在坚持自己的道路的同时，受到同行太多的反对。奥坦涅米的迪波里学生活动中心是他最好的作品之一。后来我们在建筑设计竞赛中同他竞争过。他在坦佩雷图书馆和总统官邸

world-wide trends.

Steel and glass have been the materials we have been working with in recent years. To evaluate your own work is too difficult, it is a job for others to do.

12

What do you think of Eliel Saarinen, Alvar Aalto, and Reima Pietilä? Can you make some comments on their main works?

They were all great architects on national level and very well-known internationally as well.

Eliel Saarinen's Helsinki Railway Station is very familiar to me because we were lucky to design and build the platform roofing for this station. In my opinion, that building is still one of his best works. Eliel Saarinen went through many trends in architectural thinking starting with national-romantic style and ending up in international style.

As we all know, Alvar Aalto was the key figure in Finnish architecture from the twenties until the seventies. He started with classical style but later designed functionalistic buildings in Turku and other places in Finland. As I was born in Turku and worked there until 1969 so Alvar Aalto's work is very familiar to me. Aalto later moved in his own direction developing many great works of architecture. His humanistic approach has always fascinated me. Vuokseniska Church and Finlandia Hall are two masterpieces which I find exceptionally intriguing. Villa Mairea is a great synthesis of traditional and modern ideas in a private house and thus it belongs to the great small houses in the world.

Reima Pietilä was born in Turku although he never had an opportunity to design anything there. He was a dissident in Finnish architecture; following his own paths he was too often

15

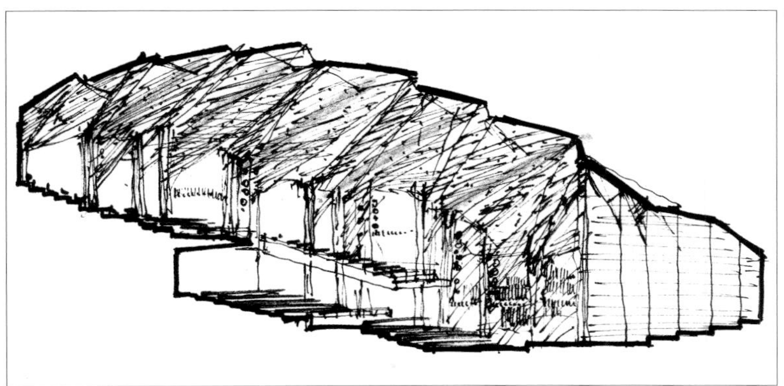
16

竞赛中都击败了我们，我们在那两次分别获得入围奖和二等奖。昆谱拉大学设计竞赛中我们赢了他，与另外两个参赛者并列一等奖。雷马·皮耶蒂莱的建筑与我们的大相径庭，这就是他为什么得不到许多芬兰建筑师承认的原因所在。我想某一天他的建筑会得到更多的承认，因为在他的建筑中我可以看到一些生态建筑的思想。

13

你是否认为生态建筑是当前最重要的任务？你如何看待建筑中的极少主义？

在今天的建筑领域，生态建筑是最富挑战性的工作。如何处理自然和人类的关系是我们面临的一个关键问题。我想我们需要从科学中获得更多的创新和思路，这样我们就有能力设计出可持续发展的建筑。在未来IT技术也许会帮助我们，但是我们必须记住人类在过去的很长岁月里都是非常相似的，发展需要时间。

建筑中的极少主义是实现生态建筑的一种方法。巴克敏斯特·富勒常常问道，一幢建筑可以承受多少。极少主义并不是最终的解决之道，因为它留下了太多没有解答的问题，而且正如罗伯特·文丘里所说："少就是枯燥。"

14

当你设计一座建筑的时候，是否会考虑它与城市以及建筑室内的关系？

rejected by his colleagues. The Student House Dipoli at Otaniemi is one his best works.

Later we were competing with him in architectural competitions. He beat us in Tampere Library competition, where we won a purchase prize and in the Official Residence for the President , where we won the second prize. We beat him in Kumpula University competition, where we won the first prize divided among two other entries. Reima Pietilä's architecture was different from what we were used to and that is why he was not respected by many architects in Finland. I think his architecture will get more respect some day, because in his ideas of ecological architecture can be seen.

13

Do you regard ecological architecture as the most important task of today? How do you think of Minimalism in architecture?

Ecological architecture is one of the most challenging tasks in the field of architecture today. How we are dealing with nature and human beings is a key issue for us. I think we need more innovations and ideas from science so that we are able to design sustainable buildings. IT-technology may help us in the future, but we have to remember that human beings have been quite similar for a long period in the past and development takes its time.

Minimalism in architecture is one way to do research in ecological architecture. Buckminster Fuller always wondered how much a building can weigh. Minimalism is not however the final solution, because it leaves so many questions unanswered, and as Robert Venturi pointed out: sometimes less is a bore.

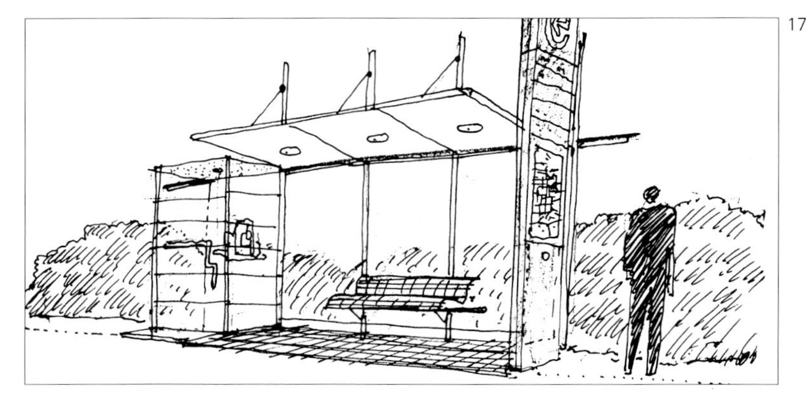

17

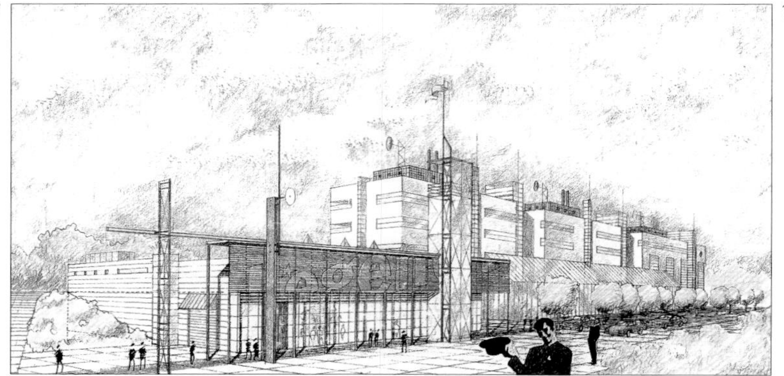

18

17 公共汽车站，埃斯波，1990年
18 办公大楼Atrium，埃斯波，1990年方案
19 Tikkurila Orthodox教堂，凡塔，1991年竞赛入围
20 办公大楼Tiistinportti，埃斯波，1991年方案

17 Bus shelter, Espoo 1990
18 Office building Atrium, Espoo, project 1990
19 Tikkurila Orthodox Church, Vantaa, competition entry 1991
20 Office building Tiistinportti, Espoo, project 1991

你亲自设计室内还是让室内设计师去做？

我想，环境属于我们大家，这也是为什么建筑师（受过专业训练以从事这一行的专门人员）至少可以设计建筑的外部环境，甚至桥梁和其他构筑物的原因所在。一座建筑的室内，如果是属于私人的，那么或多或少该由业主自己去处理。公共建筑属于大家。有时候我们会请室内设计师以及景观设计师、声学专家和其他专业人员加盟我们的工程。

只有室外和室内同时得到设计的建筑才是最好的建筑。

15

你如何与工程师合作？他们什么时候参与工程？我感觉近十年来对工程师这一角色越来越重视了。

建筑是一项团队工作，每一个参与者均拥有自己的分工和专业知识。如果没有工程师，我们的建筑很难建成。工程师们都很现实，而且他们知道房子如何才能盖起来。所有的专业人员都很早就加入我们的工程，因为我们想了解设计工作的极限和可能性。有时候第一位专家是一名景观设计师或者声学顾问。结构工程师总是与我们的设计组紧密合作。几年前工程师的角色比一个建筑师还要重要，但我认为我们采取的做法是正确的：建筑师对我们的环境负有主要的职责，是整个乐队的指挥。

14

When you design a building, how do you consider the relationship between the exterior and interior of the building? Do you make the interior or have interior architects to complete it?

I think the environment belongs to all of us and that is why architects (specialists who have been educated for that) should design at least the exterior of buildings and even bridges and other major structures. The interior of a building, if it is a private one, is more or less for the owner to deal with. Public buildings belong to all of us. Sometimes we have had interior designers participate in our projects as well as landscape architects, acoustics and other specialists.

The best buildings are created when the exterior and interior are designed simultaneously.

15

How do you collaborate with engineers? When do they enter the project? In my opinion there has been greater recognition of the role of engineers over the past ten years.

Architecture is teamwork where every participant has his own role and special knowledge.

Engineers are realistic and they know how the buildings are built. All the specialists start in our projects at an early stage because we want to know the limits and possibilities of the design work. Sometimes the first specialist will be a landscape architect or an acoustical consultant. A structural engineer is often working very closely with our team. Some years ago the role of an engineer was greater than that of an architect, but I think we are going to the right direction where an architect has the main

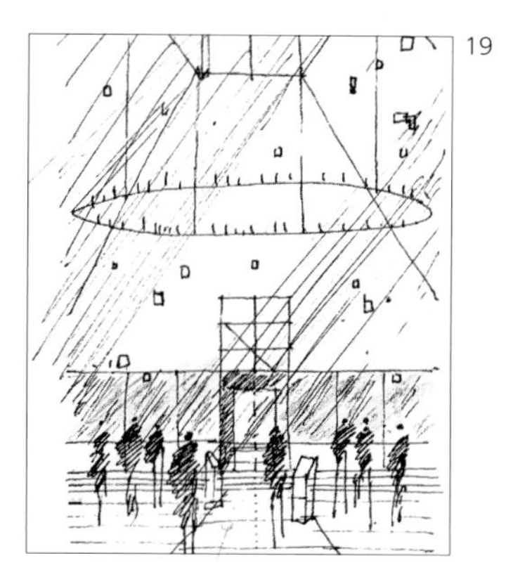

19

20

16

你对建筑理论和历史的课题感兴趣吗？你是否认为理论与设计实践之间有着紧密的关系？你认为历史是一个百宝箱吗？

当我在Otaniemi技术大学教书时，我也得为自己的建筑技术副博士学位而从事研究工作。我对环境心理学特别有兴趣，因为我仍然认为建筑为人而造，知道人们对自己的环境有何反映是至关重要的。我的副博士论文是“人对环境的反映测试”，而且我至今仍在研究这个领域。

建筑历史是一个百宝箱。所有的构思在某时某处就早已以某种方式存在了，这话我早就说过。我们的任务是创造出新的星座和可能的方式，以解决现代问题。我们必须为人类设计遮蔽物，满足他们的需要，也许有时会让他们惊奇诧异。

17

在对设计意象的探讨和材料问题之间是否存在着一种联系？在过去所建的许多工程中，主要依赖混凝土，也许部分原因在于其形式。但近来似乎采用新材料、新技术模式等手段的可能性得到了拓展……

在建筑中，没有一种单独的材料优于其他材料。我们的任务始终是为建筑的某个特定的局部或细节找到最好的材料。有些材料是热的，有些是冷的，每一种材料都拥有自己的特性，我们必须决定在什么地方使用木材，什么地方用钢、混凝土等等。

responsibility of our environment and is the conductor of the whole orchestra.

16
Are you interested in the theory and history of architecture? Do you think theory has a close relationship with design practice? Do you think history is a chest of treasures?

When I was teaching at the Technical University at Otaniemi I had to do research for my licentiate of technology degree. I was especially interested in environmental psychology because I still think that buildings are especially designed for human beings and it is very important to know how people react to their environment. My licentiate thesis was called "Measuring Human Responses to the Environment" and I am still studying that area.

The history of architecture is indeed a chest of treasures. As I said before almost all ideas already exist somewhere. Our task is to produce new constellations and maybe new ways to solve modern problems; we have to design shelters for human beings, to satisfy their needs and maybe even touch them sometimes.

17
Is there a link between the exploration of design ideas and the question of materials? In many of the projects built in the past years, there is a reliance on concrete, maybe partly because of the possibilities it offers. But recently there seems to be an increasing amount of new materials, new techniques of prototyping and so on.

There is no single material that is better than others in

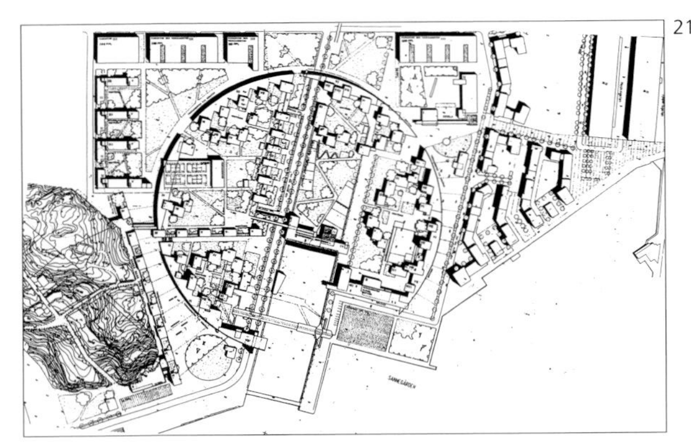
21

22

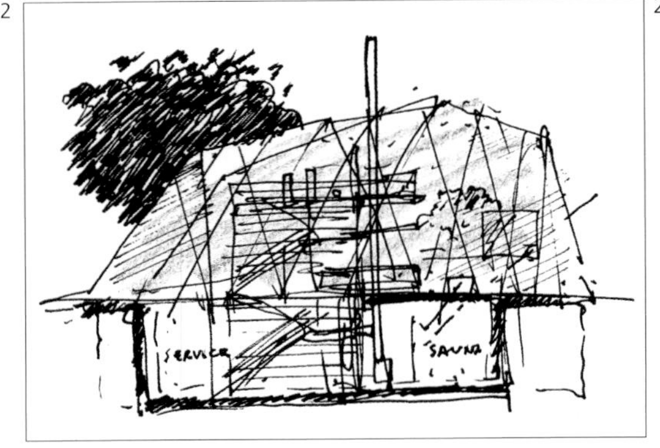

23

21 Västra Eriksberg住宅区，哥特伯格，1992年竞赛入围

22 赫尔辛基火车站候车站台屋顶，1994年

23 玻璃住宅，1996年方案

24 Rautaruukki Polska技术中心，波兰2000年

25 Leppävaara中转站，埃斯波2001年

21 Västra Eriksberg Housing Area, Göteborg, competition entry 1992
22 Helsinki Railway Station Platform Roofing 1994
23 Glass House, project 1996
24 Rautaruukki Polska Technology Center, Poland 2000
25 Leppävaara Exchange Terminal, Espoo 2001

在60年代芬兰开始了这样一个时期：大建筑公司纷纷投资混凝土建材厂，同时我们也必须在较大的城市中非常迅速地建造大量的公寓以容纳来自乡间的新住户。其他材料被放在一边，那是一个很大的错误。现在我们在不同材料之间找到了一个好得多的平衡。钢和玻璃在办公和交通建筑中用得很多。木材用于小住宅以及人们希望触摸的建筑当中。混凝土越来越多只作为结构材料，其外表覆以其他材料，例如陶瓦、金属板等等。

另外建筑师当然希望尝试织物、塑料等新材料。

18

你的工作首先考虑谁？是建筑的使用者，公众，还是建筑师？

建筑师要为后代负责。建筑师最主要的顾客应该是使用者，而不是承包商、投资商和别的什么人。我们为社会和环境服务。至于别的建筑师对我们的工作怎么想并不重要。

19

在做一个建筑师的同时，你同时担任图形设计师和展场设计师，你是否认为作为一个建筑师从事一些其他领域的工作是很重要的？

当我在一家建筑事务所工作的时候，就开始做图形设计，一直贯穿了我的学生生涯。那时我也同时设计展场。我认为一个建筑师的知识领

architecture. Our task is to always find the best material for a certain part or detail of a building. Some materials are hot, some are cool; every material has its own characteristics and we have to decide where to use wood, steel, concrete and so on.

In the sixties there started a period in Finland when big construction firms invested in concrete element factories and we had to build very quickly lots of apartments in bigger cities for people that moved from the countryside. Other materials were rejected and that was a great mistake. Nowadays we have a much better balance with different materials. Steel and glass are used a lot in office buildings and transportation buildings. Wood is used in small houses and in places people want to touch the material. Concrete is more and more used as a structural material and clad for instance with ceramic tiles, metal panels and so on.

And of course architects always want to try new materials like textile, plastics etc.

18

To whom is your work primarily directed to: the users of the building, the public or the architects?

Architects are responsible to future generations. The user should be the main client of an architect not the contractor, investor or other people. We serve the society and environment. What other architects think of our work is of minor importance.

19

You are an architect, but you are also a graphic designer as well as an exhibition designer. Do you think it is good to do some works in other fields as an architect?

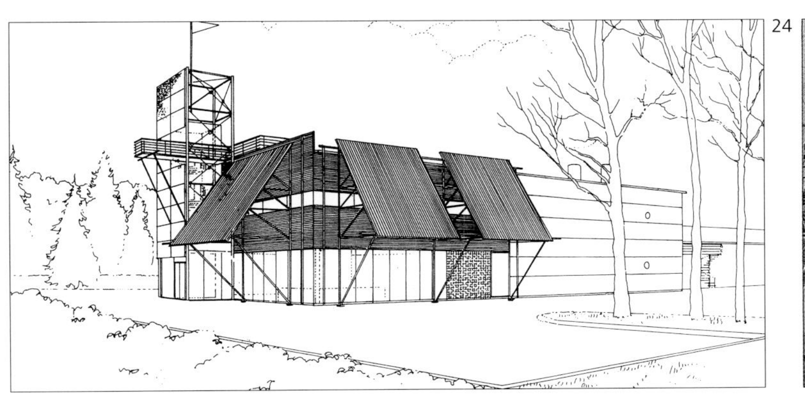

24

25

26 Koivukylä 火车站，万塔，2002 年

27 Mäntsälä 火车站，2003 年

域应该尽可能的宽广。我们应该成为通才而非知识面狭窄的专家。当然如果你成了一个百事通却一无所长，那也不是一件好事。

20

你有没有自己喜爱的当代建筑师？你是否愿意对一些重要建筑师的作品作些评论，诸如史蒂文·霍尔、阿尔瓦罗·西扎、扎哈·哈迪德、雷姆·库哈斯、让·努韦尔、弗兰克·盖里、伦佐·皮亚诺、理查德·罗杰斯和诺曼·福斯特？

我对我们当代建筑师的工作非常感兴趣。所有的历史时期都有许多权威出现。有些明星会比其他星星更亮，但到了下一个时期就消失了。研究和评价上述建筑师是很有意思的，但因为处于同一时代，要对他们的作品说出什么深刻的话未免为时太早。

现在高技建筑在全球广受欢迎。因此福斯特、努韦尔、皮亚诺和罗杰斯对我来说比较熟悉。我曾经参观过他们的许多建筑，但我至今认为他们的建筑中缺少某些人文关怀。他们的建筑干净，冰冷，细部精美，多少有些索然寡味。也许他们正在开创第二种国际式。

弗兰克·盖里的生平和建筑对我有很大的吸引力。我认为他的毕尔巴鄂（Bilbao）古根海姆博物馆是北半球最出色的新建筑之一。

26 Koivukylä Railway Station, Vantaa 2002
27 Mäntsälä Railway Station 2003

I started with graphic design when working at an architectural office as a student. I also designed exhibitions at that time. I think the area of knowledge for an architect should be as broad as possible. We should become generalists but not narrow-minded specialists. But of course becoming a jack-of-all-trades and master of none is not a good thing.

20
Do you have your favorite contemporary architects? Would you like to make some comments on the works of some prominent architects such as Steven Holl, Alvaro Siza, Zaha Hadid, Rem Koolhas, Jean Nouvel, Frank Gehry, Renzo Piano, Richard Rogers and Norman Foster.

Naturally, I am very interested in what our contemporary architects are doing. There are many new gurus appearing all the time. Some stars glitter more than others, but in the next moment they may disappear. All the above mentioned architects are very interesting to study , but at the same time I think it is too early to give a final evaluation of their works.

High-tech architecture is very popular now all over the world. So Foster, Nouvel, Piano and Rogers are familiar to me. I have seen many of their buildings, but I think some human touch is still missing in them. Their buildings are clean, cool, very well detailed, somewhat sterile; maybe they are starting the second international style.

Frank Gehry's career and buildings fascinate me quite a lot. I think his Guggenheim Museum in Bilbao is one the most remarkable new buildings in the northern hemisphere.

26

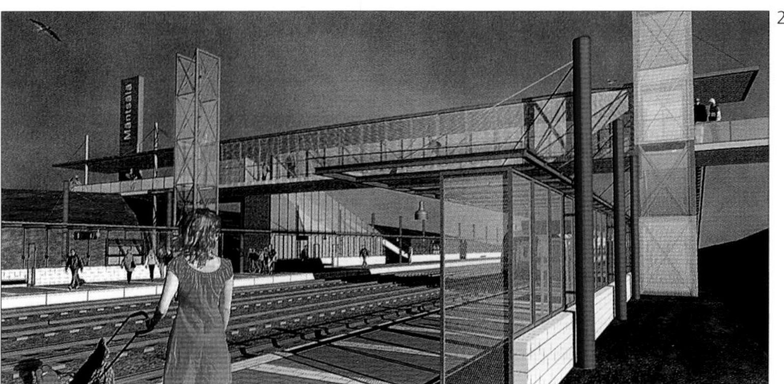

27

建筑作品 Architecture

私人游泳馆，图尔库 1967

Private Swimming Hall, Turku 1967

这个游泳池满足两方面要求：首先，它是一个蓄水池，储存经氧化过的深井水，从而提高了花园的产量；它同时也是一个配有桑拿和休息区的游泳池。这个游泳池可以常年使用。

建筑物是由钢柱和钢梁支撑的。整个结构被焊接在一起。涂成红色的钢材和朴素的绿玻璃构成了整个正面。颜色的选择表现了建筑物本身的材料：钢和玻璃。而这样的颜色的搭配贯穿在整个建筑当中。

桑拿室裸露的红砖墙、门和窗框的色彩、休息区的家具，这一切营造出一个和谐的氛围。

The pool fulfils two needs: it functions as a reservoir in which deep-well water is oxidized and thus improves the market garden yield, and it also serves as a swimming pool with a sauna and relaxation area. The pool can be used all the year round.

The building is supported by pillars made of steel sections as are the joists.

The structures have been welded together. Red-painted steel and simple green glass dominate the façade. The colours were chosen as symbols of the building materials: steel and glass. The colour scale has been implemented consistently throughout the building.

The bare red brick walls of the sauna, the colours of the window frames and doors and the furniture of the relaxation area all add to the harmonious overall appearance.

1 室内景观
2 桑那
3 角部细部
4 细部
5 立面
6 立面和剖面
7 从起居室远望
8 平面
9 室内泳池

1 Interior view
2 Sauna
3 Corner detail
4 Detail
5 Facade
6 Facade and section
7 View from the living area
8 Floor plan
9 Swimming pool

1

2

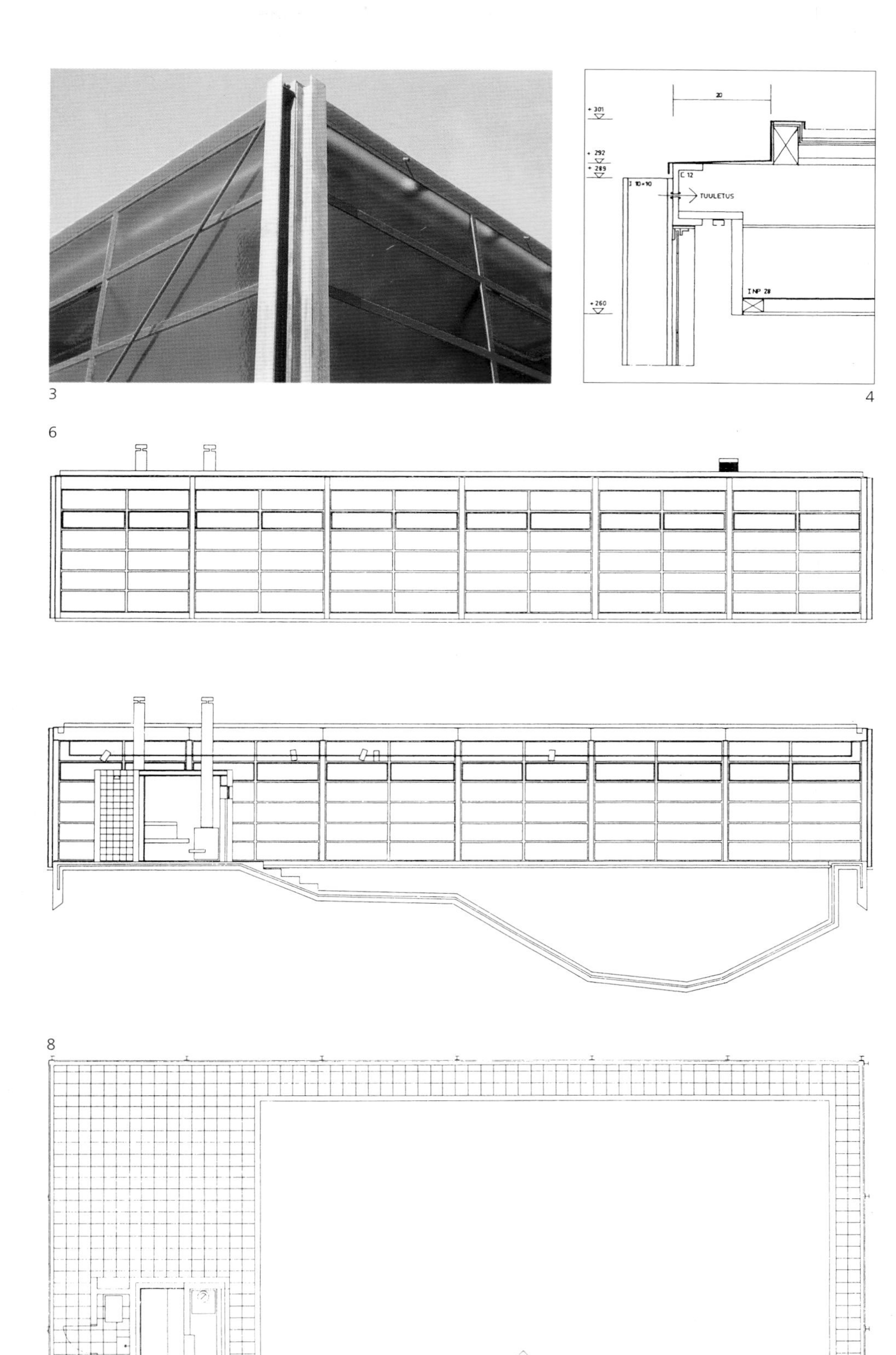

3

4

6

8

5

7

索加兰塔别墅，梅里马斯库 1969

Villa Suojaranta, Merimasku 1969

这个避暑别墅位于图尔库群岛。它是一个为四人小家庭设计的造价较低的别墅。

房子建造于朝海的斜坡上，基础由木柱支撑。

整个建筑是一个开放的空间，中间部分设有壁炉和小厨房。

卧室由家具和轻质门分隔开。

外墙则是玻璃。

The summer house is situated in Turku Archipelago. It was designed for a family of four at reasonably low costs.

The house was built on a slope towards the sea and stands on wood pillars.

The house is an open space with a fireplace and kitchenette in the middle.

The bedrooms are separated by furniture and light doors.

The outer walls are made of glass.

1

5

2

3

4

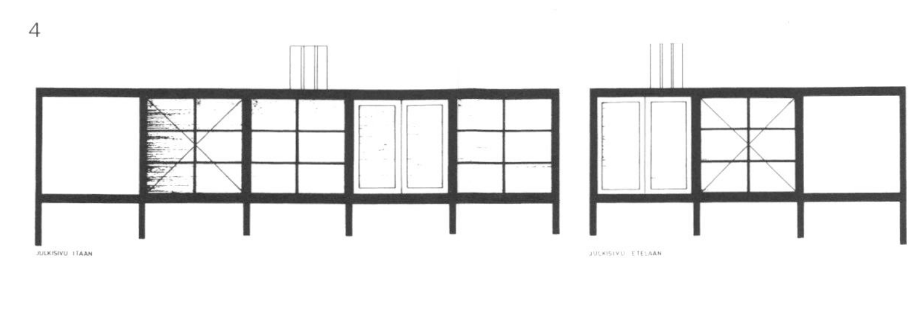

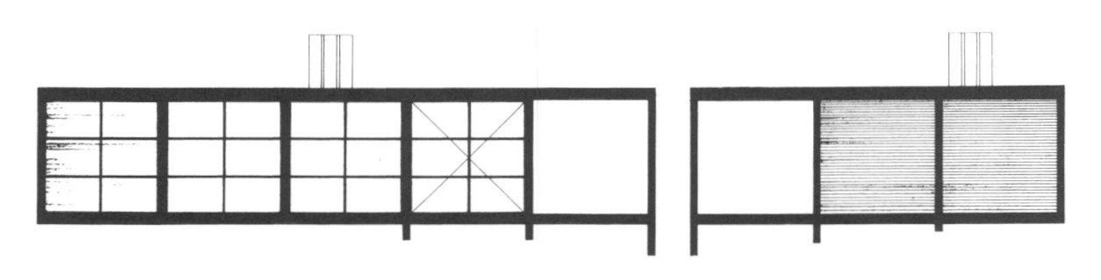

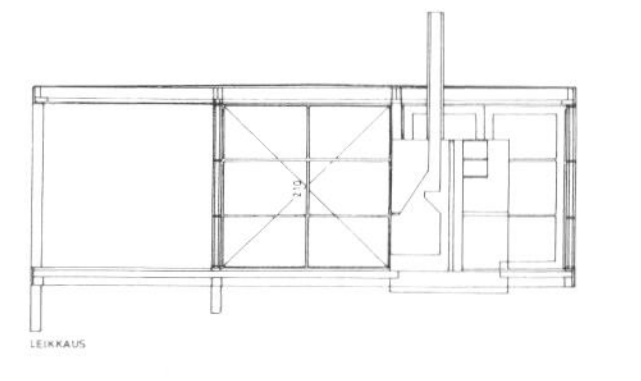

1　西立面
2　带百叶窗的西立面
3　平面
4　立面和剖面
5　窗细部

1 View of the west facade
2 West facade closed with blinds
3 Floor plan
4 Facades and section
5 Window detail

瓦伊诺拉别墅，奥拉 1970

Villa Vainola, Aura 1970

别墅位于奥拉河畔，面向河流及对岸尚未修建的河堤。

朝南和朝西的大窗户将室内和周围的环境联系在一起。

别墅为一个四口人家设计，有一个起居室／厨房，两间卧室，一个桑拿室和一个储藏室。桑拿室的更衣间也可作为一个卧室来使用。

外墙使用了白色的木料和玻璃。

建筑物配备有齐全的现代化设施，因此即使冬天的周末也可以使用。

The villa is situated on the bank of the River Aura with a view towards the river and the opposite unbuilt bank.

Large windows to the south and west connect the interior to the surrounding nature.

The house was designed for a family of four including a living room/kitchen, two bedrooms, a sauna and a storeroom. The sauna dressing room can be used as a bed room as well.

White wood and glass were used in the outer walls.

The building has all modern conveniences and can thus be used on winter weekends as well.

1

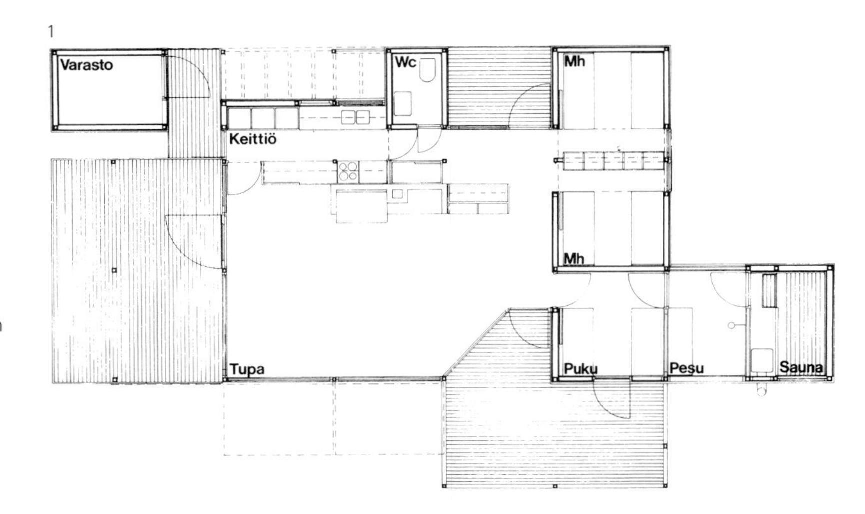

2

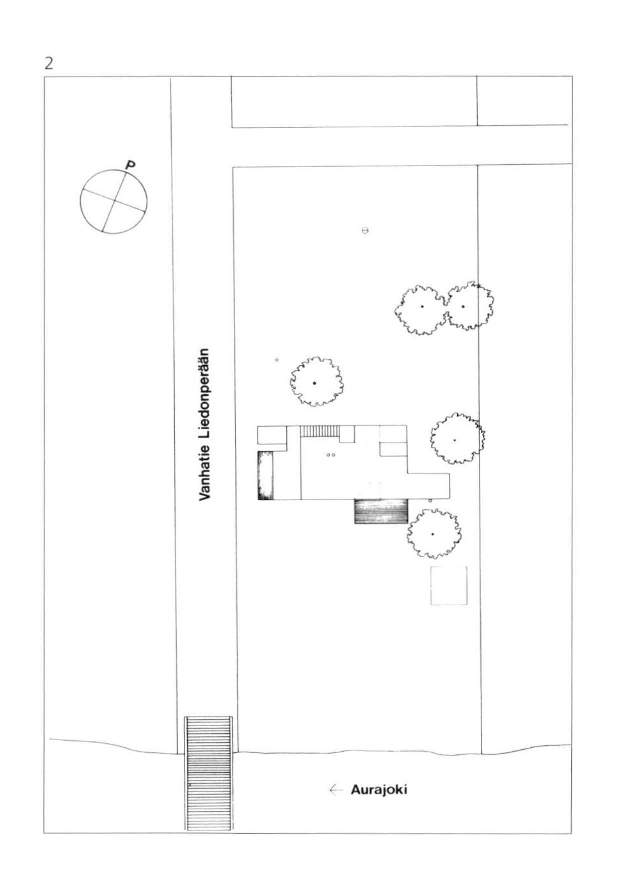

3

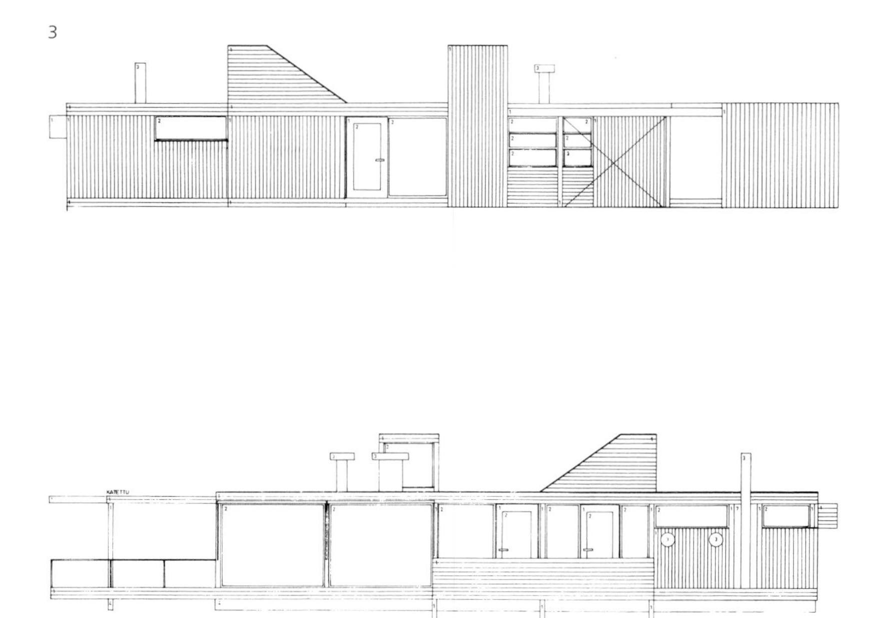

1 平面
2 总平面
3 立面
4 轴测图
5 北部景观
6 剖面
7 南部景观

1 Floor plan
2 Site plan
3 Facades
4 Axonometric
5 View from the north
6 Section
7 View from the south

4

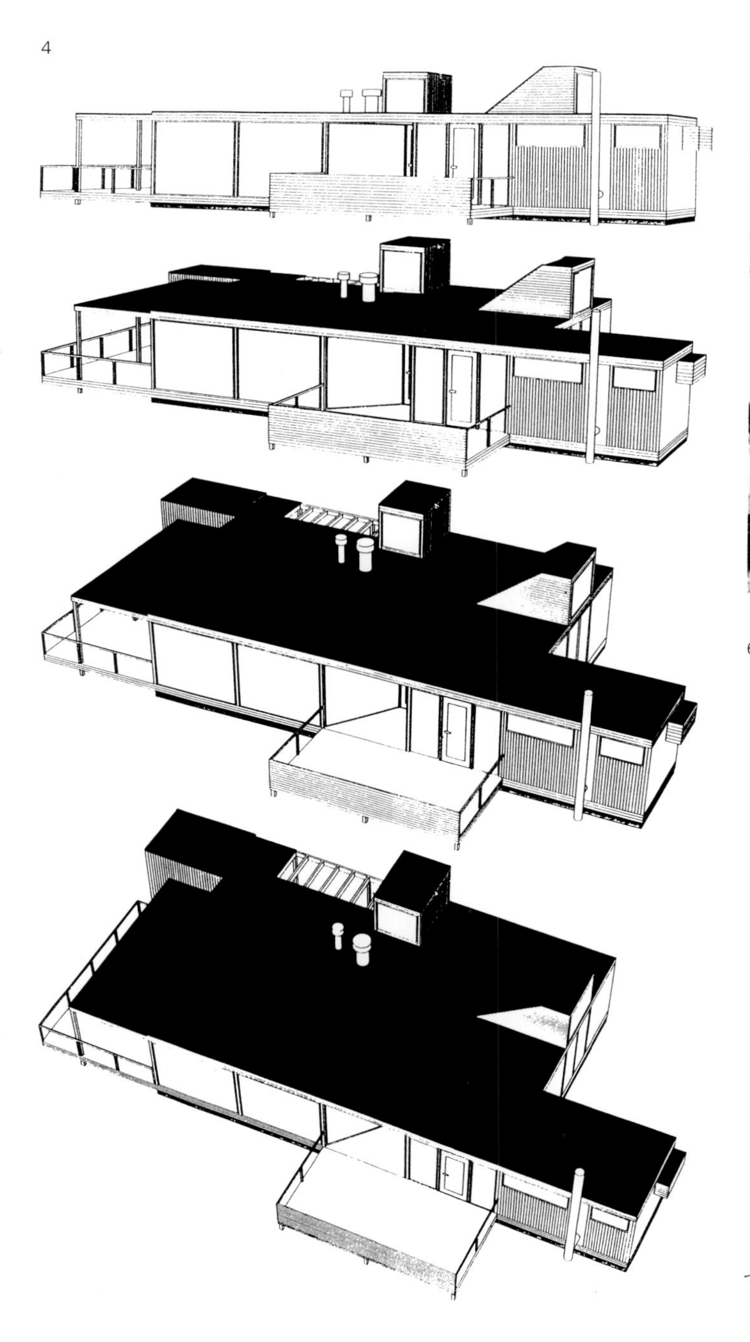

5

6

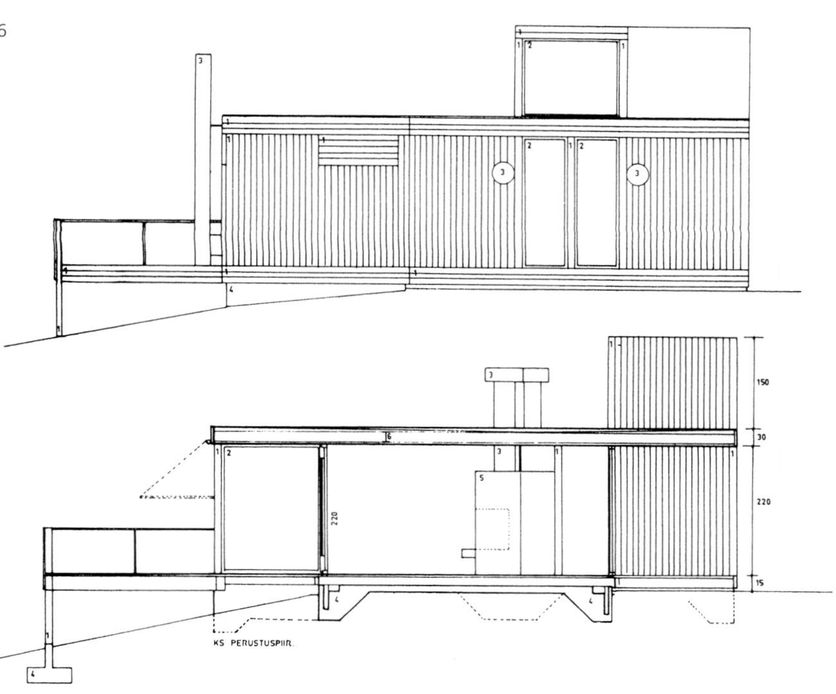

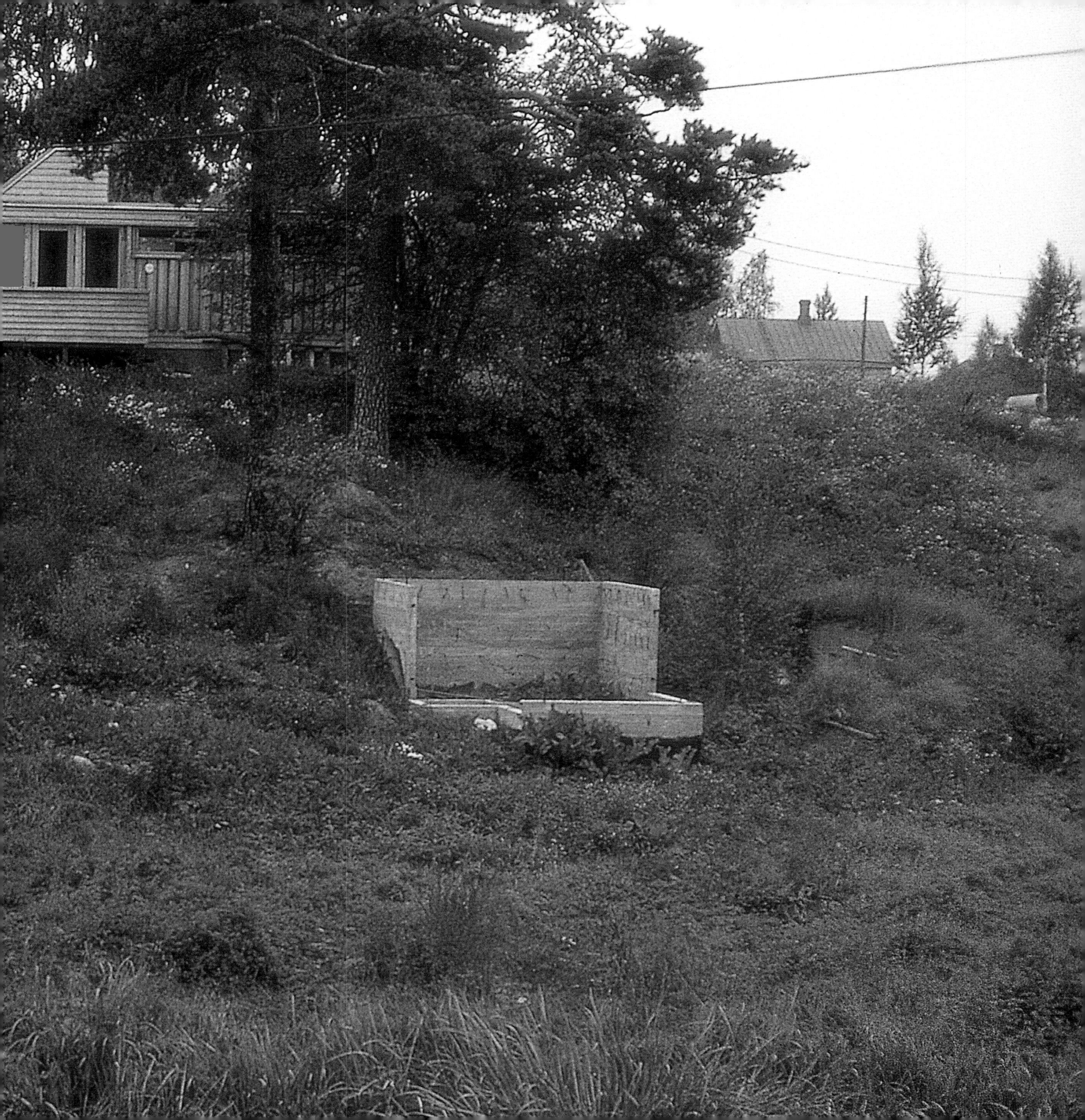

波洛周末小别墅，韦尔库 1972

Weekend Cottage Pöllö, Velkua 1972

1 立面
2 总平面
3 平面
4 剖面
5 外观

1 Facade
2 Site plan
3 Floor plan
4 Section
5 Exterior View

5

　　这是一个位于图尔库群岛中一个小岛上的建筑物，从海上几乎看不到它。小屋的构件在大陆建造，然后运到岛上组装。绿色和蓝色使得整个建筑物融入了周围的环境。

Situated on a small island in Turku Archipelago the building is scarcely visible from the sea. The cottage was built from elements built on the mainland and reassembled on the site. The green and blue colours help the building blend with the surroundings.

1

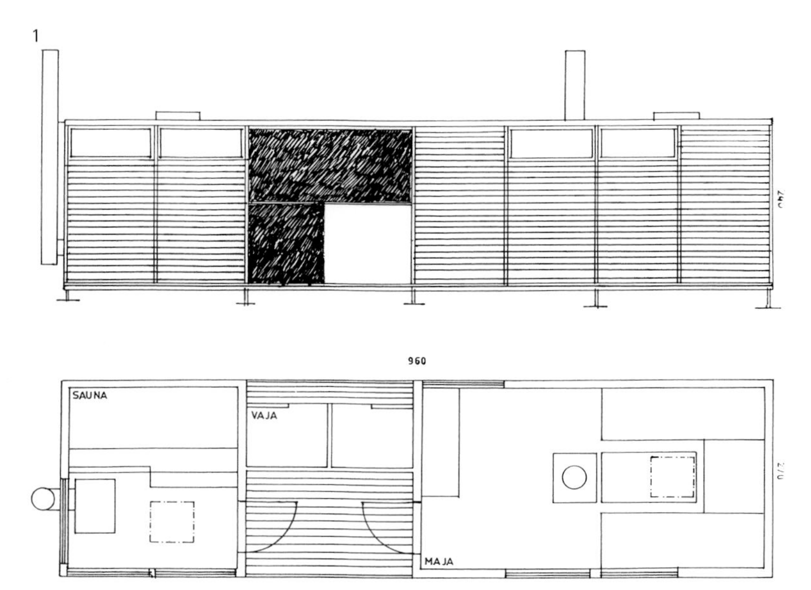

3

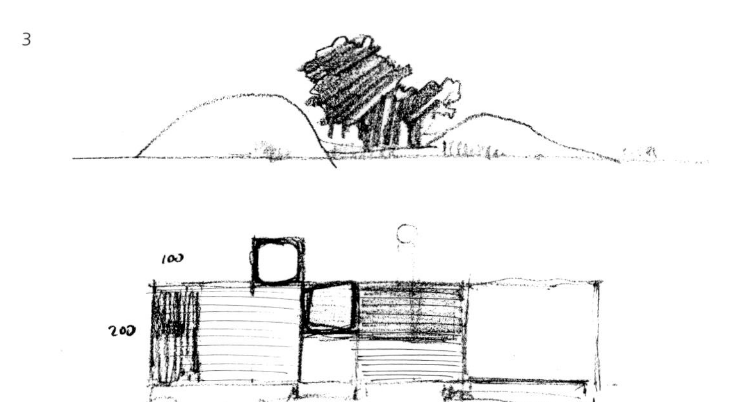

2

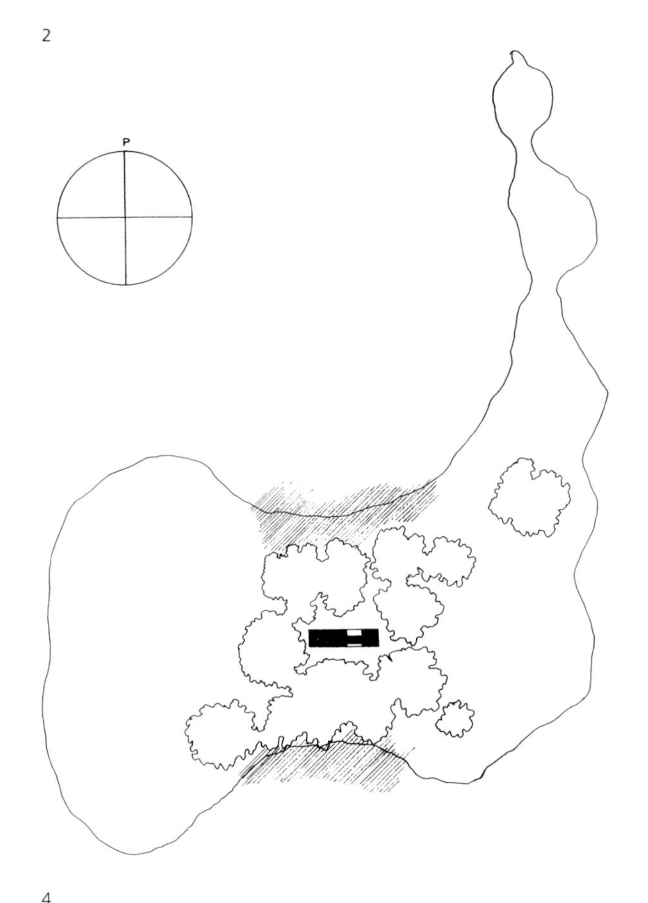

4

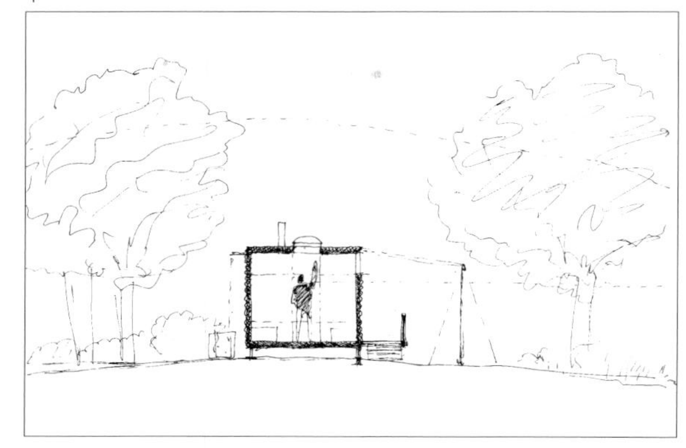

6 细部
7 天窗
8 从山坡上远望
9 立面
10 从海上远望

6 Detail
7 Skylight
8 View from the hill
9 Facade
10 View from the sea

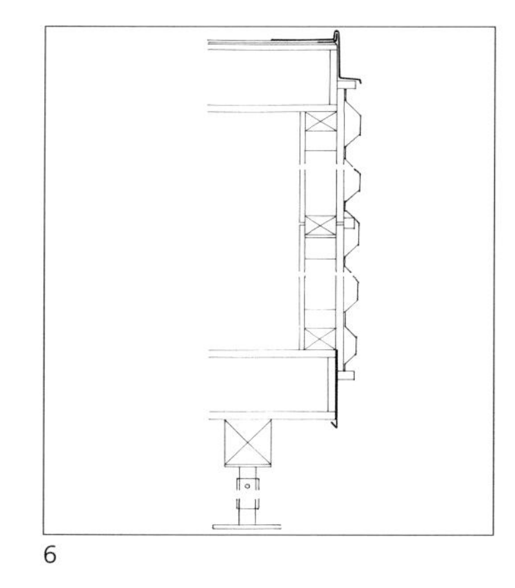
6

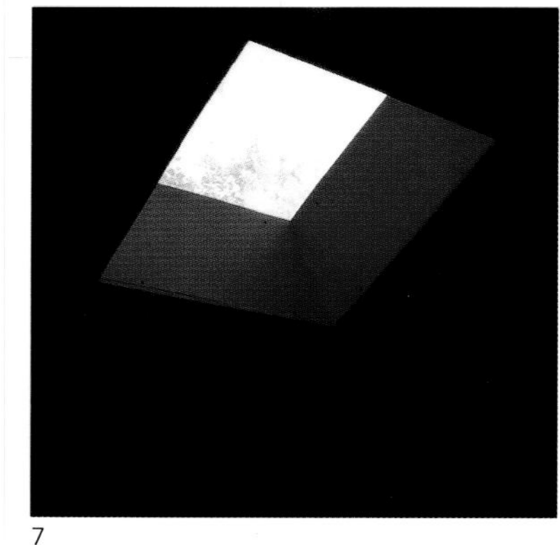
7

8

9

10

比约克霍门别墅扩建，帕拉嫩 1974

Villa Björköholmen Extension, Parainen 1974

1 从海上远望
2 剖面
3 立面
4 面海立面
5 平面

1 View from the sea
2 Section
3 Facade
4 Facade towards the sea
5 Floor plan

原别墅建于1957年，因为家庭人数的增加而需要扩建。

它原有一个外阳台和一个坡屋顶的阁楼。在改建中把阳台划到了起居室，而阁楼的一部分则改为父母亲的卧室。

翠绿色的粉刷使得别墅融入了周围的环境。

The original villa was built in 1957. The need for an extension arose with the growth of the family.

The building had an outdoor veranda and a high attic formed by the ridge roof.

In the renovation the living room was extended to the veranda and part of the attic was made into a sleeping loft for the parents.

The building was painted forest green to blend in with the surroundings.

1

2

3

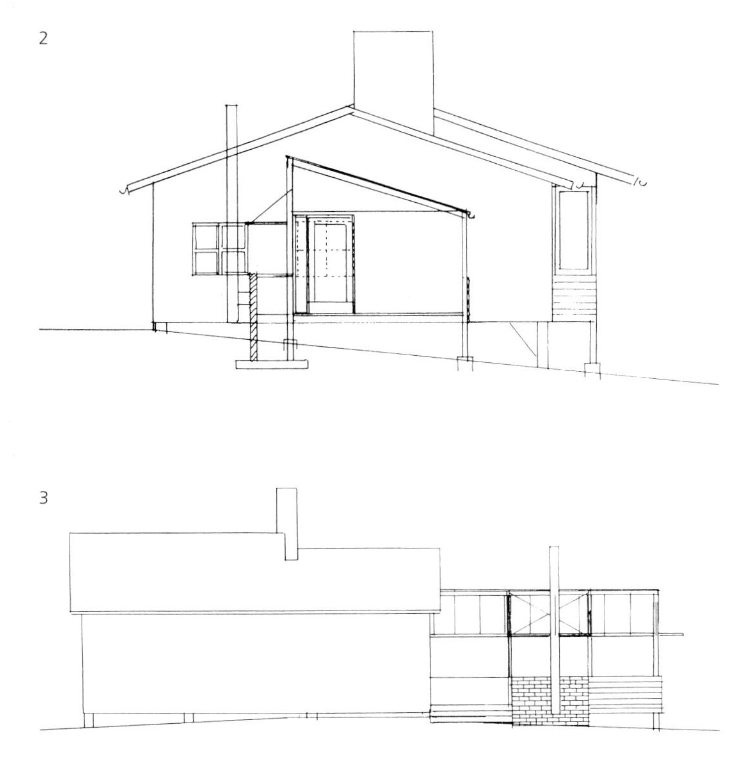

4

5

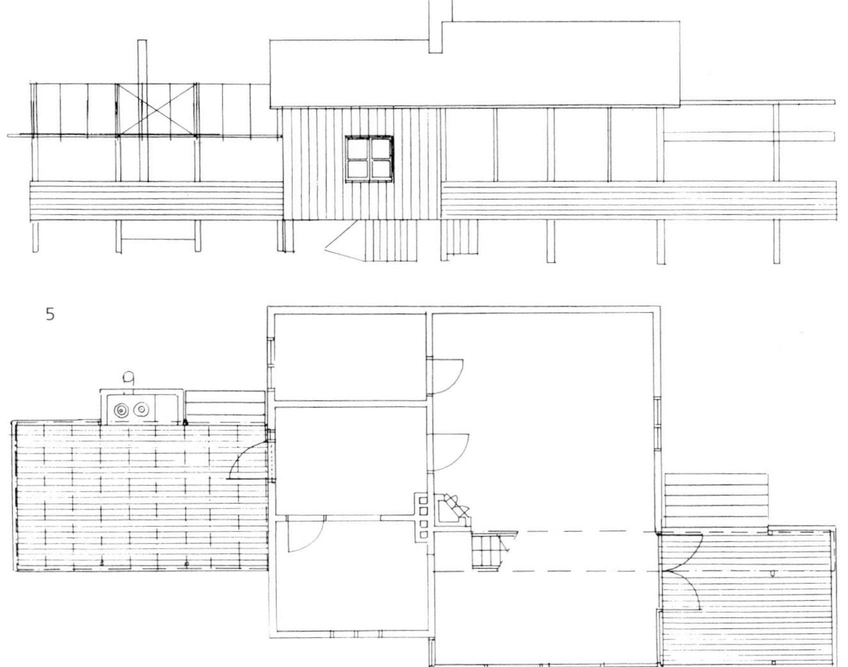

6

7

8

6 阳台
7 室内景观
8 台阶（平台）
9 平台细部

6 Balcony
7 Interior view
8 Terrace
9 Detail of terrace

皮海约基教堂，设计竞赛第一名 1975

Pyhäjoki Church, Competition entry 1st prize 1975

新教堂的入口设在被烧毁的旧教堂的废墟前。这样做的目的是为了强调原来场所的性格、强调教堂在林阴道终端的视觉焦点地位、并且为处于新教堂和森林之间的旧教堂废墟中的一个室外教堂提供遮护。

操作设备室位于教堂和教区大会堂之间，以方便资源共享。宿舍是一座有围墙的独立的建筑，位于教区大会堂对面。

建筑物和周围的树木一起，形成一个私密的教堂内院。内院为前往教堂礼拜的人提供了一个聚会场所。

教堂主体采用十字型平面。

教堂的纪念性性格表现在建筑中央宽敞高耸的空间中，这部分空间为教堂提供光线，它的体量从很远的海上就能看见，夜晚灯亮时尤其突出。

教堂和其他建筑为木结构。受力部分为复合木材和实木面板。

In the entry the church was placed in front of the ruins of the burned-down church. The aim was to emphasize the previous character of the place, the church as a visual dominant at the end of an avenue of trees, and to shelter the outdoor church situated within the ruins of the old chuch, between the new church and the forest.

The operating facilities have been placed between the church and the parish hall, so that flexible parallel use is possible. The apartment is a building in its own surrounded by a fence opposite the parish hall.

Thus the buildings together with the trees form an intimate church yard that serves as an assembly place for churchgoers.

The church itself is cruciform.

The sacral character of the church has been emphasized by a high spacious central area providing light to the church and which can be seen far from the sea, especially when lit.

The church as well as the other buildings are built of wood; the supporting structures are glue-laminated wood with wood weatherboarding.

1

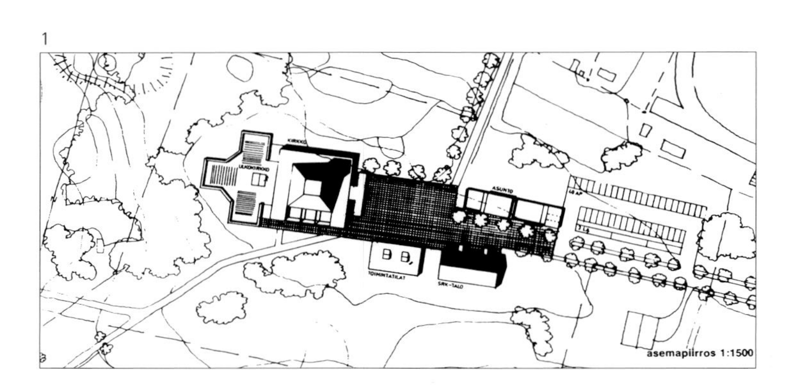

2

3

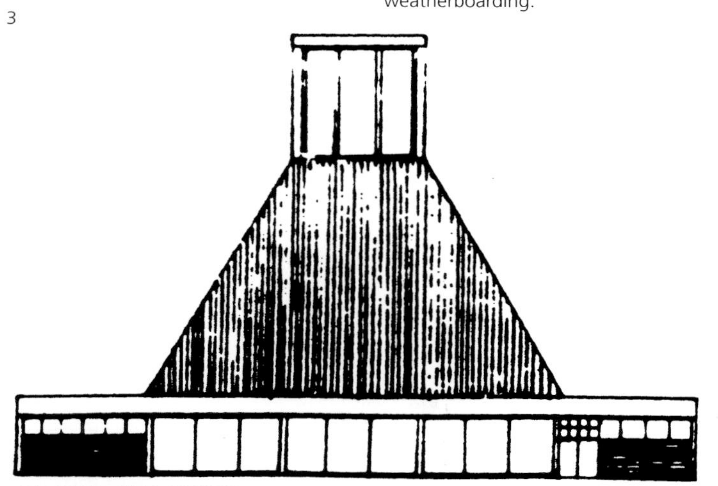

4

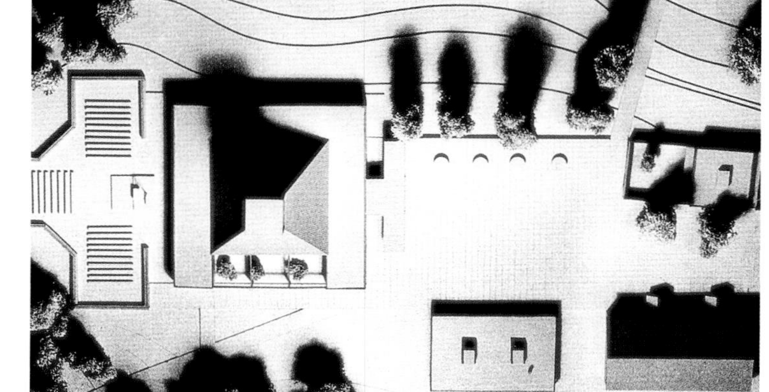

5

6

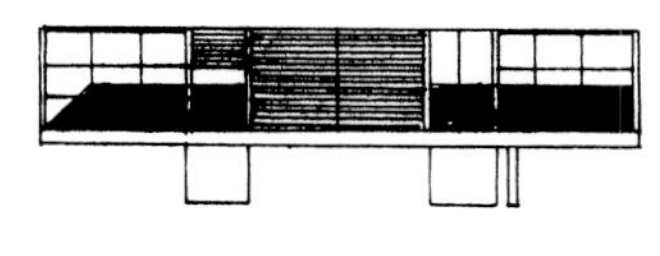

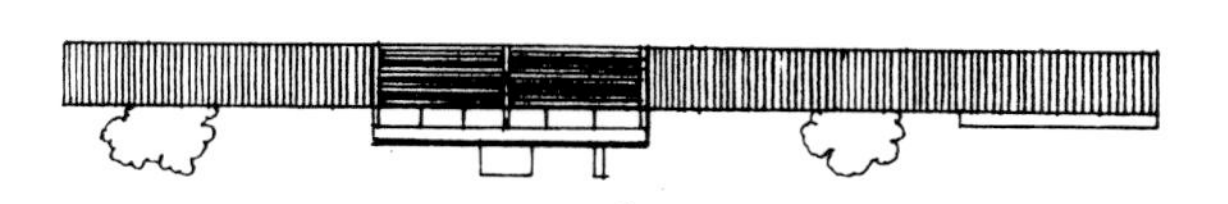

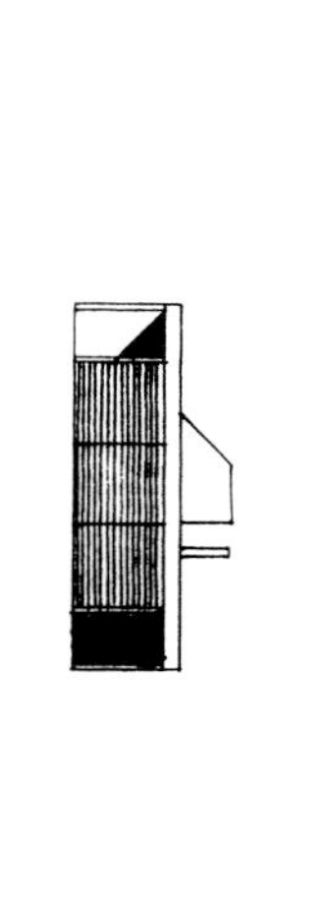

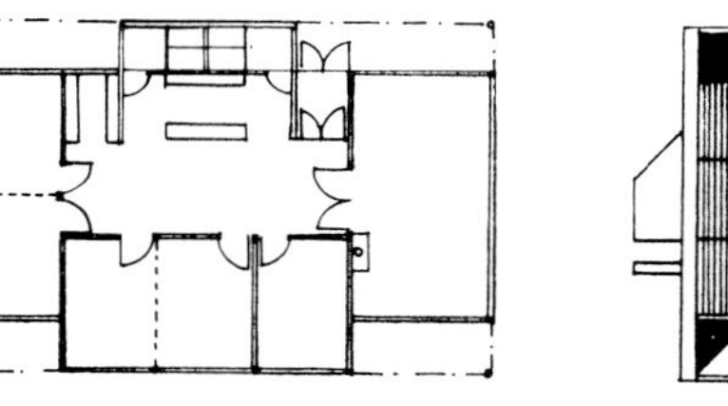

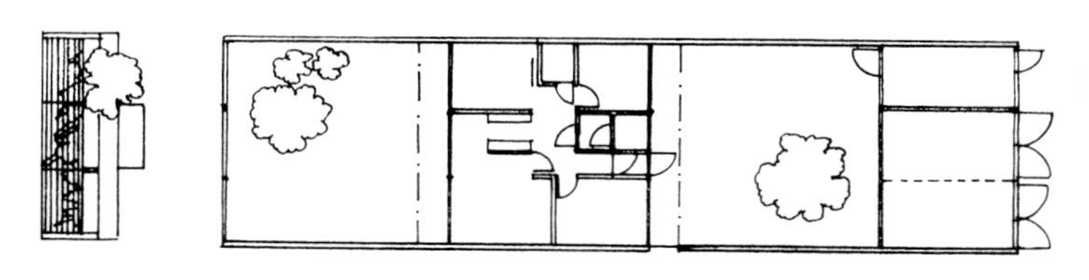

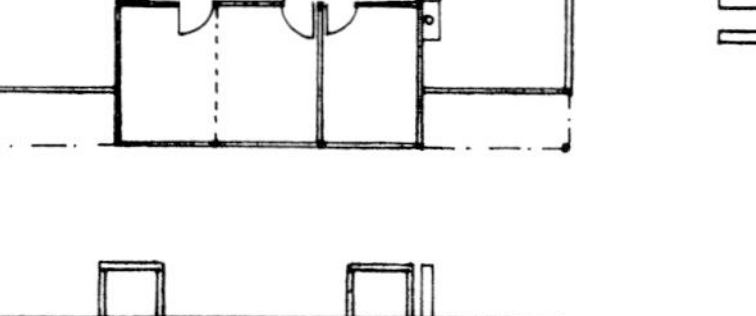

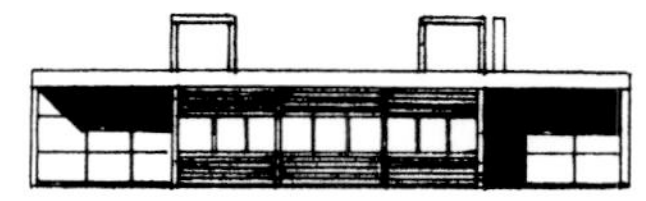

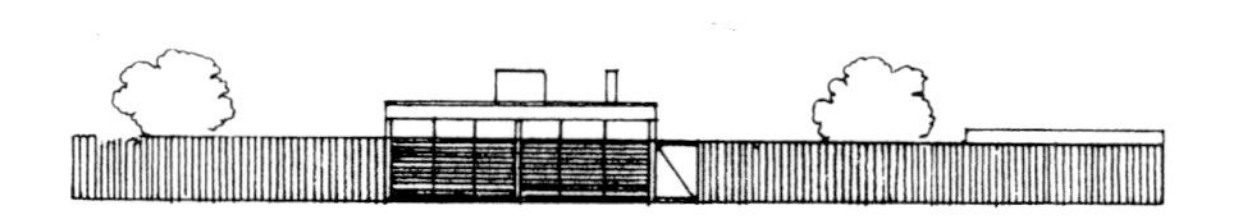

1 总平面
2 平面
3 北立面
4 模型照片
5 模型照片
6 平面和立面
7 外观透视
8 室内透视

1 Site plan
2 Floor plan
3 North facade
4 Model photo
5 Model photo
6 Floor plans and facades
7 Exterior perspective
8 Interior perspective

7

8

昆普拉大学地区，设计竞赛第一名（并列），赫尔辛基 1979

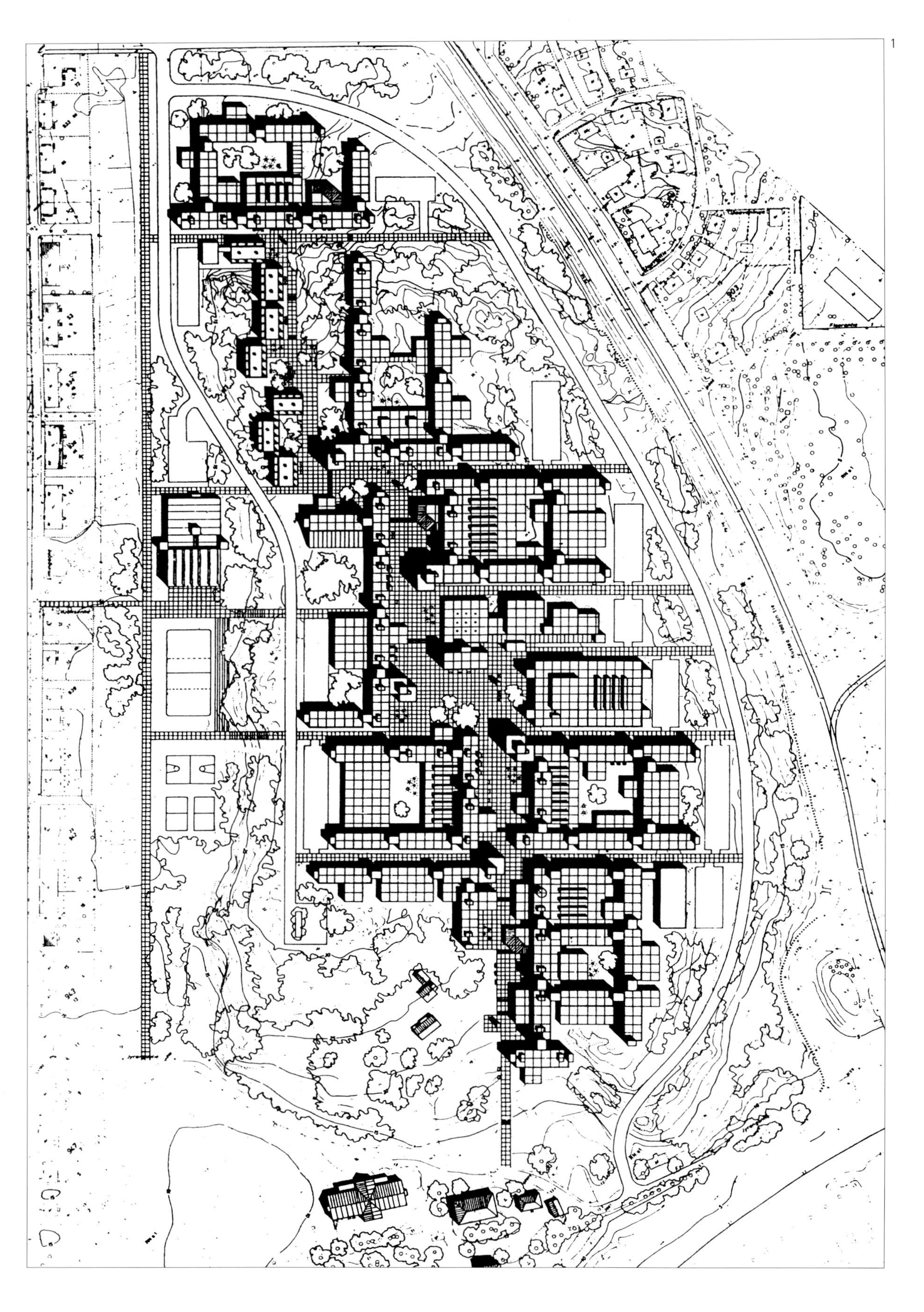

由教育部、赫尔辛基大学建筑委员会和赫尔辛基市联合主办了一个关于昆普拉大学地区及其周边环境城市设计的方案竞赛。

竞赛的目的是为该地区，特别是赫尔辛基大学数学和自然科学系系馆区域作远期规划，并为建筑发展提供指导性建议。规划目标是为了形成高质量、多功能、同时造价经济的城市结构框架和城市景观。设计特别注重各类城市功能之间的交互融合。对于校园内建筑本身的设计，必须考虑到较长的建设周期，并且能够适应不可预见的使用功能上的改变。除大学外，还需为紧邻校园的一个植物园确定位置，并为其制定功能、环境和景观发展规划。

尽管当地的建筑传统是以低层为主，但由于建设地点位于一座小山上，因此从城市里依然可以看到这群建筑的天际线。

我们在这里建议运用的建构原则是：以主体结构和附属的功能元素（实验室、楼梯塔）相结合的方式控制建筑的整体形式。垂直的主体不但加强了建筑群的特点同时减小了尺度感。

1 总平面
2 模型照片
3 鸟瞰剖面
4 鸟瞰剖面

1 Site plan
2 Model photo
3 Aerial section
4 Aerial section

Kumpula University Area, Competition entry 1st prize (shared), Helsinki 1979

The Ministry of Education/Building Board of Helsinki University and the City of Helsinki sponsored a general ideas competition for the urban structure of the Kumpula university area and its immediate surroundings.

The aim of the competition was to make a proposal to be used as the foundation for the further planning and construction of the competition area and especially the area to be used by the Faculty of Mathematics and Natural Sciences of Helsinki University. The objective was to obtain a high-quality urban structure and townscape, offering functional variety, and economical to carry out. It had to pay special attention to the integration of various urban activities. With the university buildings themselves it had to take account of the long building period and permit unpredictable changes in functions. A plan with high-quality functioning, environment and townscape also had to be proposed for the location and construction of a botanical garden immediately off the univeristy area.

The building site, a hill, despite low construction tradition, meant that the outline of the area could still be seen in the horizon of the city.

The architectonic building principle proposed: the main framework with its additional functional elements (laboratories, staircase towers) makes it possible to dominate the overall architectonic appearance: the vertical motifs both characterize the identity of the area and reduce the scale (San Gimignano).

2

3

4

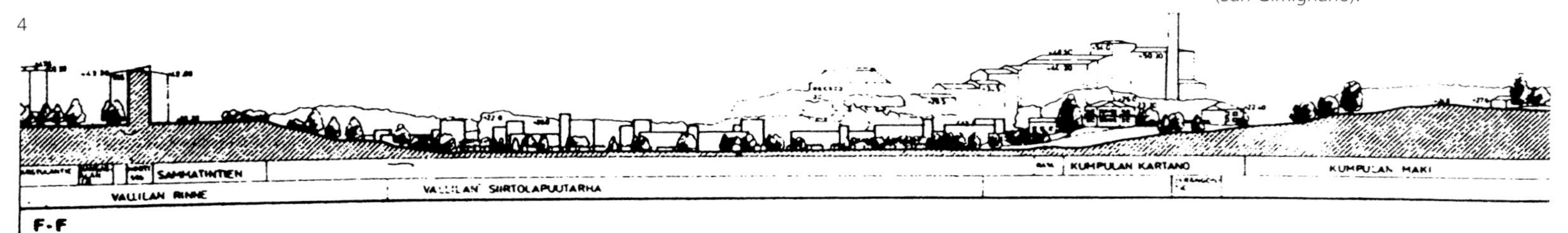

5

6

7

8

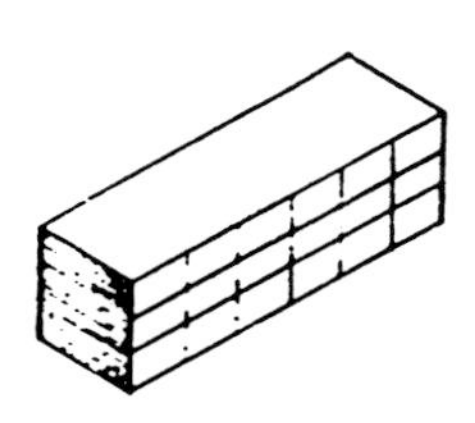

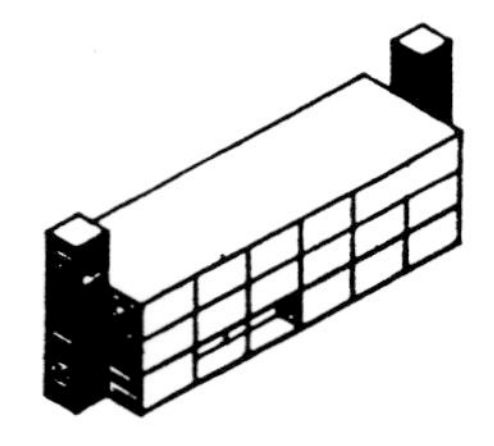

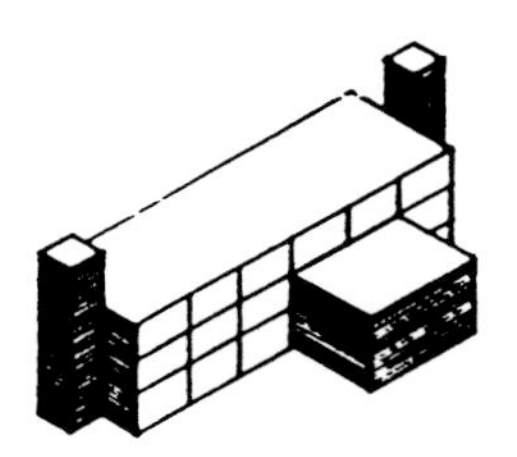

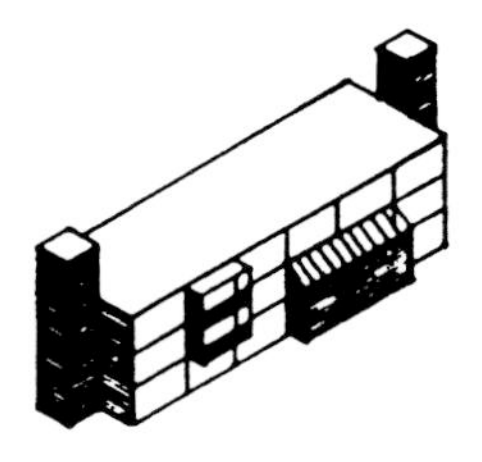

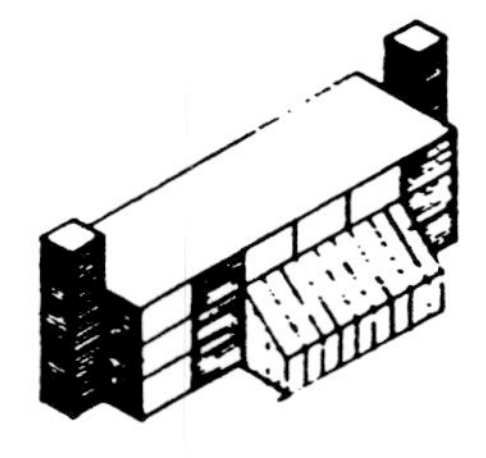

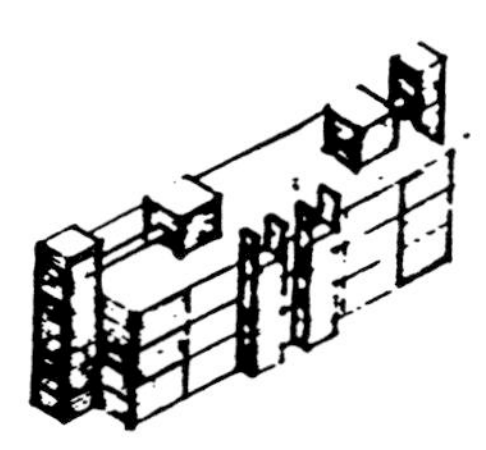

5 模型照片
6 广场视角透视
7 步行街视角
8 建筑体系

5 Model photo
6 Perspective view from the square
7 Pedestrian street view
8 Building system

劳塔瓦拉教堂，1982

Rautavaara Church, 1982

1 总平面
2 剖面
3 平面
  1 教堂室内
  2 入口大厅
  3 圣器收藏室
  4 等候室
  5 储藏室
  6 技术室

1 Site plan
2 Section
3 Floor plan
  1 Church interior
  2 Entrance hall
  3 Sacristy
  4 Waiting room
  5 Storage
  6 Technical room

1979年4月5号的夜晚，劳塔瓦拉教堂被大火烧毁，这座颇有文化价值的木教堂是由C·巴西设计，在1825—1827年间建造的。

新教堂需要有一个新的、与众不同的解决方案，因为被烧毁教堂的残垣断壁与它曾经给人们留下的强烈印像依然存在。同时，留存下来的钟楼、当地特有的小村风貌和开阔的山野风景都是新设计需要考虑的重要因素。如何在老的、民众曾经对其投入了深厚感情的建筑遗址上重新建造一座新的建筑是当今建筑学界的一个重要课题，它涉及到如何尊重传统和人们对旧建筑的怀念之情，同时又满足新时代新建筑的需求（先进的建筑技术和新的功能需要等等）。

设计的着手点是当地的环境和传统，将旧教堂的意象凝练成象征符号（如旧教堂的巴西利卡式平面）与当今的建筑技术和细部处理手段相结合。本设计的另一个目标，是用简单的技术手段、朴素而真实的材料、基于材料自然质地的色彩体系，营建出一座乡村教堂。

在建筑内外结合使用了两种最基本的技术和材料手段，将敦实的砖墙和轻盈通透的钢木结构相结合。砖墙通过凹陷（祭坛、唱诗席位置）和开洞（花龛和侧门）的处理手法加强效果；与厚重的墙体形成对比的是小块面木质的肌理和屋顶钢结构的天窗。

On the night of April 5, 1979 the culturally valuable wooden church in Rautavaara, designed by C. Bassi and built in 1825-1827, was destroyed by arson.

In designing a new church it was necessary to find new and unusual solutions, since the powerful shape and image of the burnt-down church still lingered in the landscape, and the old belfry, the small-scale village setting and the wide-open hill country all set their requirements. A situation where a new building is to be erected where an old one has been burnt down and where the destruction of the old building is invested with such emotion, is a vital element in today's debate on architecture: how to deal with the requirements of tradition and the emotional content involved and the demands of today (advanced building technology, new functional needs etc.)

The starting points were the environment and tradition, the image of the old church, tradition crystallized at the level of symbols (the basilica plan of the early church), which was to be implemented with today's technology and architectural detail. Another architectural goal was to produce a country church, using simple, uncomplicated technical solutions, genuine materials and a colour scheme based on natural materials.

The technical solution and materials combine two basic elements both inside and outside: the massive masonry is emphasized with recesses (altar, choir stalls) and openings in the walls (recesses for flowers and doors in the side walls) and with the airy wood-panelled ceiling and glass wall surfaces, with the faceted wood and steel structure of the skylight resting on the plastered walls.

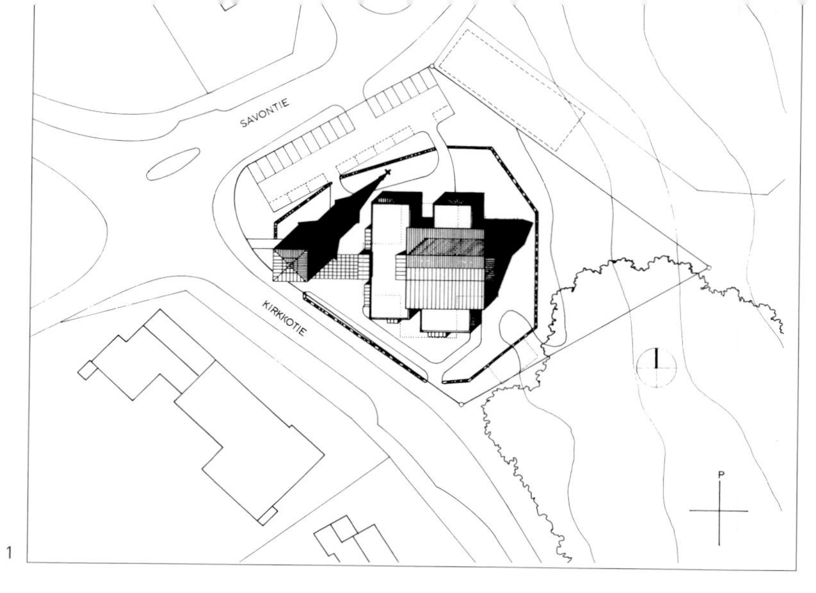

1

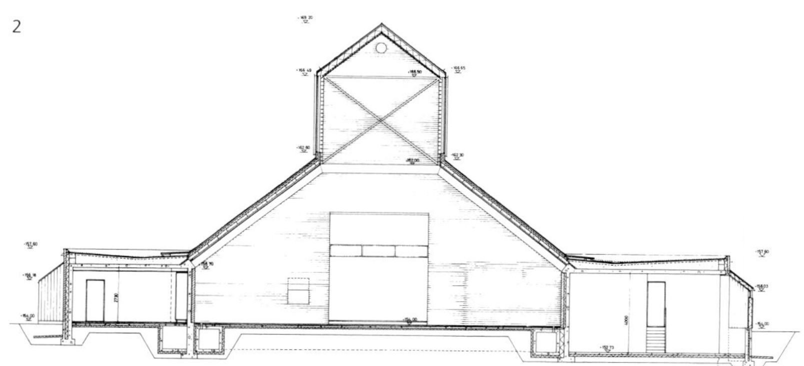
2

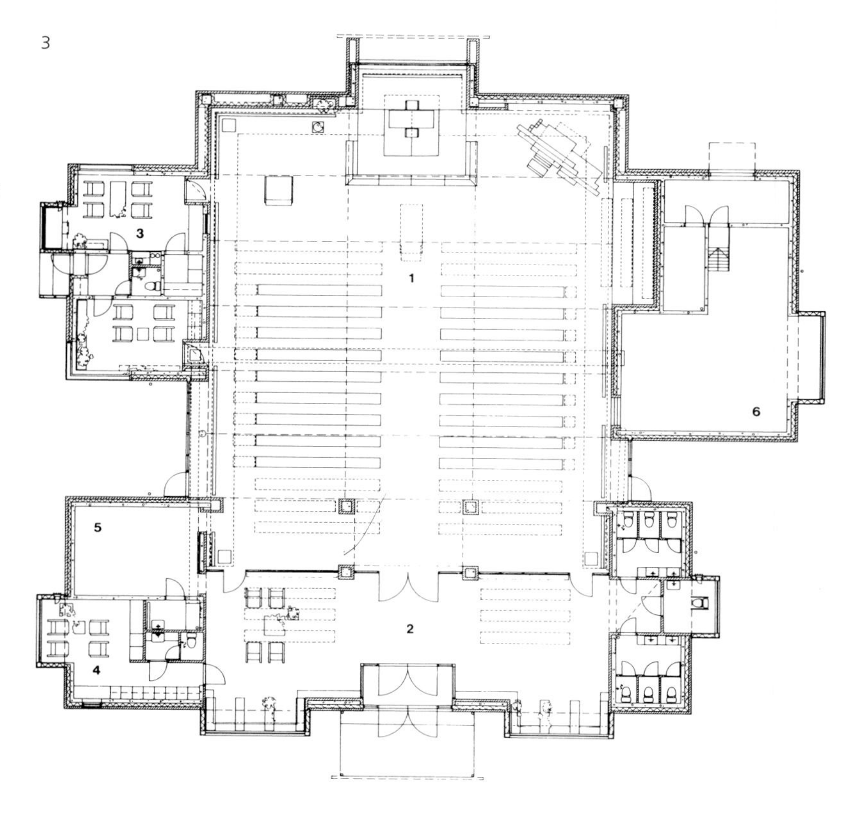

3

51 NURMES

4

4 全貌
5 立面
6 剖面
7 村落中央的教堂
8 老钟楼和新教堂
9 主要入口
10 外观

4 General view
5 Facade
6 Section
7 Church in the centre of the village
8 Old belltower and new church
9 Main entrance
10 Exterior view

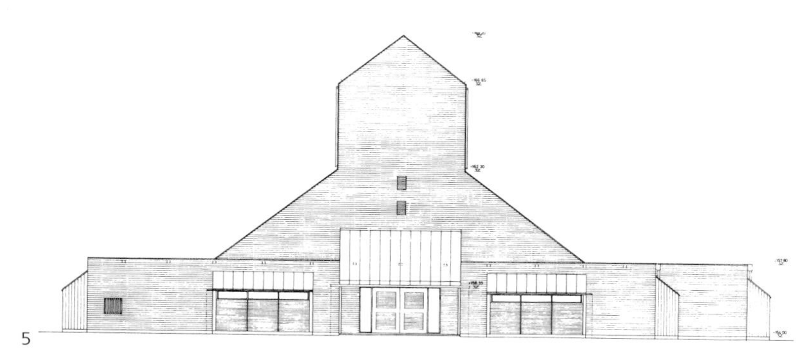
5

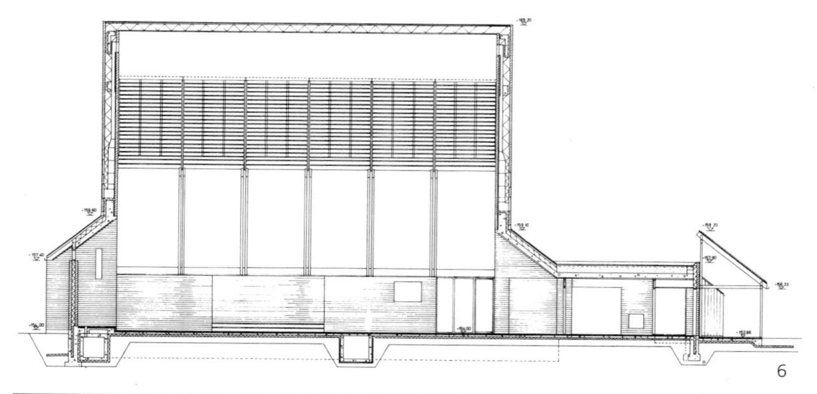
6

7

8

9

10

11

11　室内
12　从入口大厅看
13　天窗

11　Interior
12　View from the entrance hall
13　Skylight

12

13

芬兰总统官邸，设计竞赛第二名，赫尔辛基 1985

Official Residence of the President, Competition entry 2nd prize, Helsinki 1985

芬兰的总统官邸是周边环境中最引人注目的一部分。

从大门进入，一条小路曲曲折折，几棵松树围绕着一个裸露的石山，石山前树立着一个小型雕像。在小路的最高处可以看到大海，沿着一道墙可以到达僻静的入口庭院，院子的中间是一个水池，水池周围栽满了蓝白色小花，池的中央是一个喷泉。

建筑放置在地块的边缘，与原有建筑同处一边。这个手法使以前的传统、景观和视野都可以获得延续。服务性车流控制在地块的另一边。地块的大部分区域保持着自然状态，使建筑有着广阔的视野。

在功能布局上，建筑延续了以前的传统：前面是入口庭院，接待室位于主楼，服务设施和生活区位于主楼的侧翼。

在形式上，强调的是房间和空间的不同层次：接待室规整且布置传统，与比较自由的居住区和楼下分散的休憩区域相对比。维修和服务区位于后翼。

接待厅的布置强调的是文化传统的特点：即由侧面的小厅堂和餐厅围合形成的一个中央大厅。中央大厅通过凸窗获得室外自然景观。侧边的小厅与图书室相通。宿舍的布置与接待厅相对比，形成一种传统的城市公寓特有的舒适气氛。

休憩区是卡累利阿森林池塘的城市版本：一个向周边开放的空间，其中有水体、绿色植物、自然山石元素，这些元素在由接待楼建筑限定而成的长方形空间中自由布局，满足休闲需要。

建筑基础结构材料是混凝土。立面材料是浅色面砖、大理石、涂料和自然石。钢结构部分为深蓝色。

The official residence of the President of Finland is part of the surrounding landscape, featuring its highlights.

Inside the gate there is a winding path through the grounds, a bare outcrop of high rock surrounded by pine trees, with a low sculpture in front of it; the highest point of the path offering a view of the sea, a wall leading to the well- sheltered entry courtyard, in the middle of which a pool with white and blue flowers and a soothing fountain.

The building was placed at the edge of the plot, on the site of the previous building so that the tradition, landscape and views of the old site can be used.

Maintenance traffic is confined to the other edge of the plot. Most of the plot was left in its natural state leaving the building with a view over an untouched landscape.

Functionally the building continues the old manor tradition using the methods of modern architecture: an entry courtyard, reception rooms in the main building, and maintenance facilities and living quarters in the side wings of the main building.

Tha basic form of the building emphasizes the hierarchy of the rooms: the formal, traditionally laid out reception rooms contrast with the freeform living quarters and a sprawling recreation area downstairs. The maintenance and service facilities are in a rear wing.

The traditionally laid out reception rooms emphasize a cultural tradition: a large salon in the middle flanked by a smaller salon and the dining room.

The large salon views the landscape through a bay window and glass-frame greenhouse. The more intimate small salon is connected with the library.

The layout of the private apartment is in contrast to that of the reception rooms with the aim to create a relaxed and cosy atmosphere of old city

1

apartments.

The recreation area is an urban application of a Karelian forest pond: a space opening directly to the surroundings, lanscaped with water, plant and natural rock motifs, built freely within the rectangle formed by the reception rooms and designed for leisure use.

The supporting structures of the building are concrete. The elevation materials are pale ceramic tile, marble, plaster and natural rock. The steel frames are dark blue.

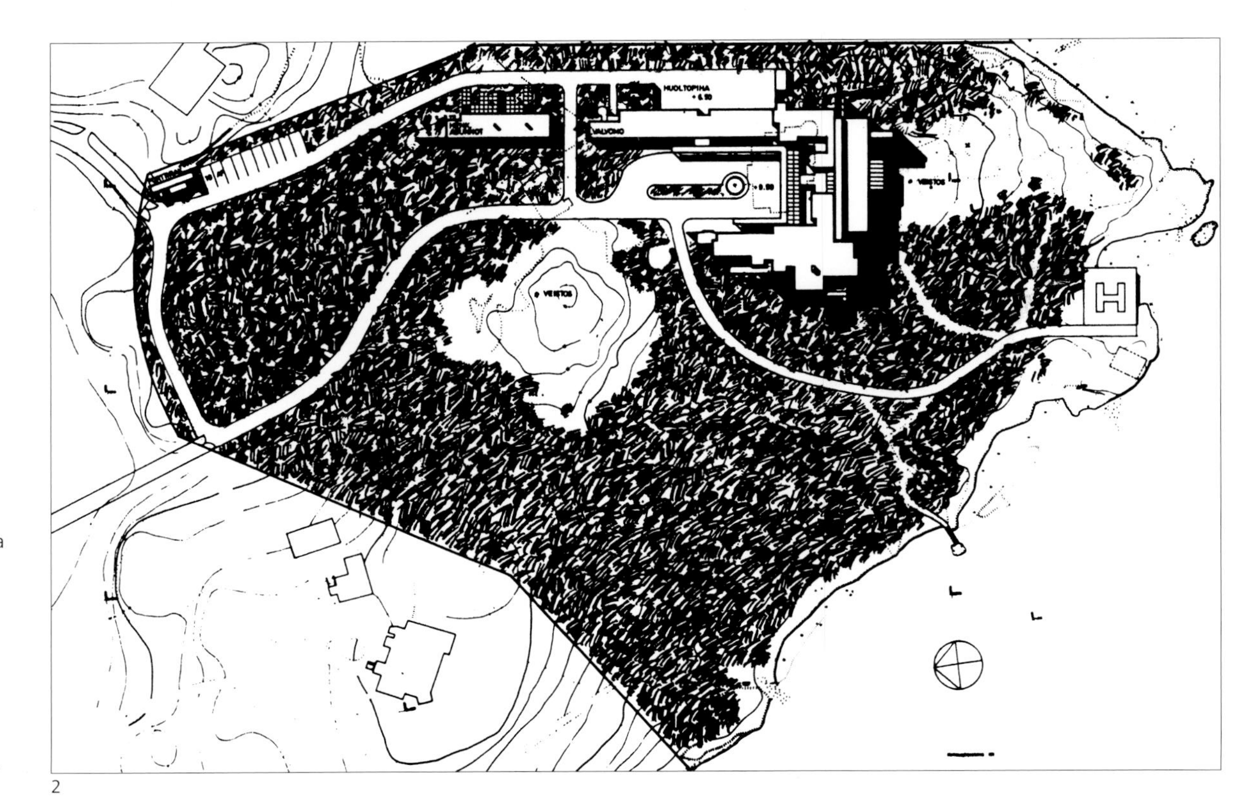

2

3

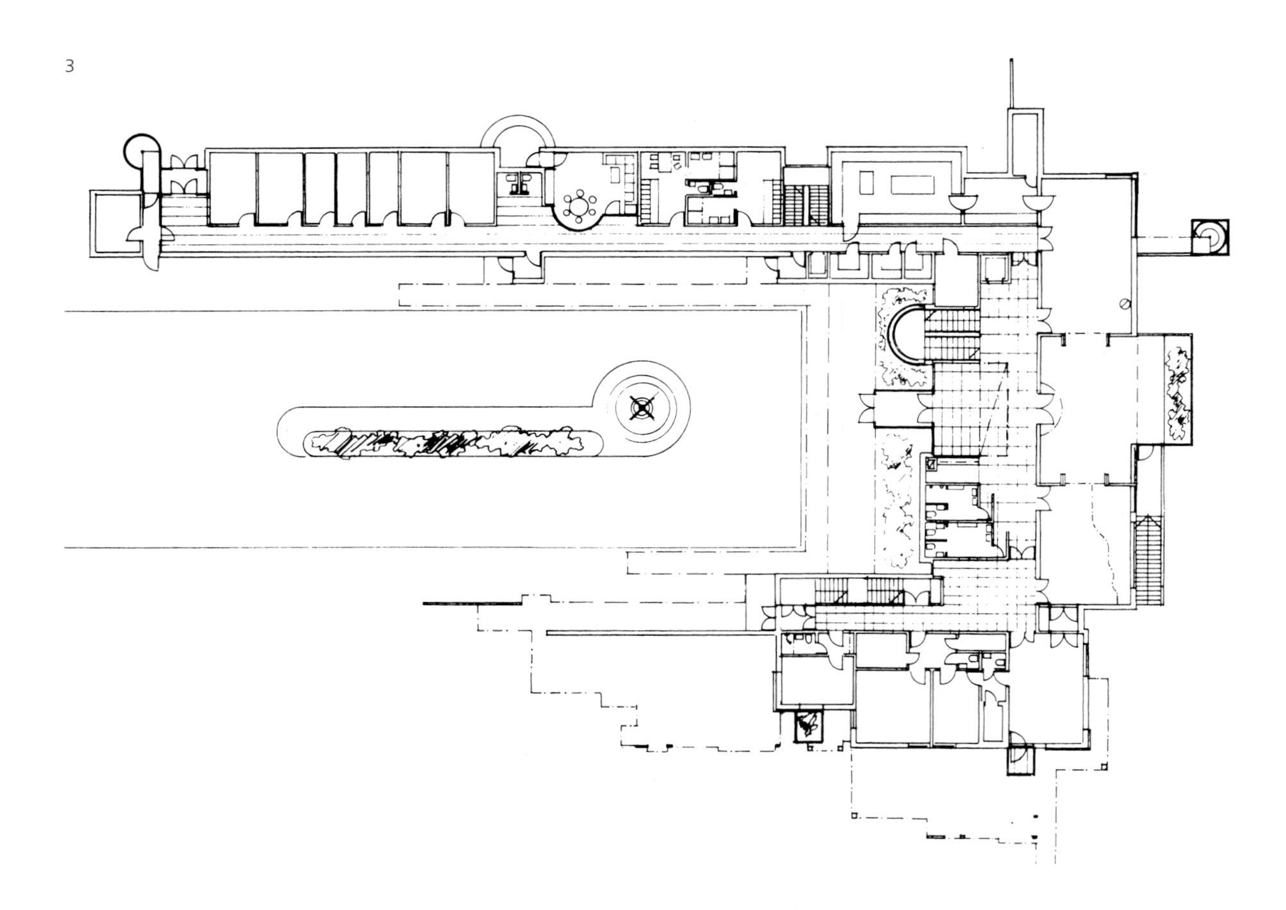

1 模型照片
2 总平面
3 平面
4 西立面
5 东立面
6 剖面
7 剖面
8 南立面
9 室内透视
10 北立面

1 Model photo
2 Site plan
3 Floor plan
4 West facade
5 East facade
6 Section
7 Section
8 South facade
9 Interior perspective
10 North facade

4

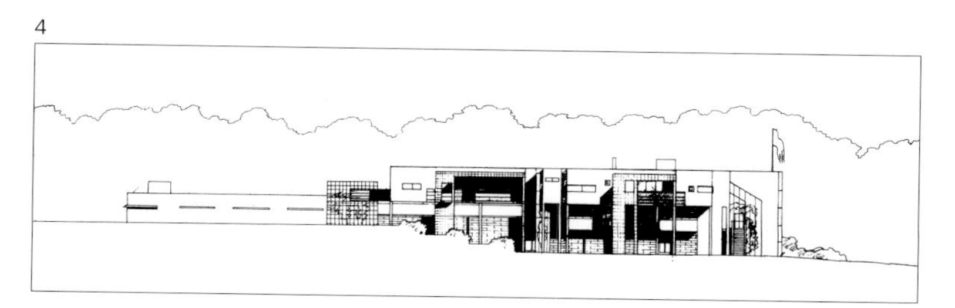

5

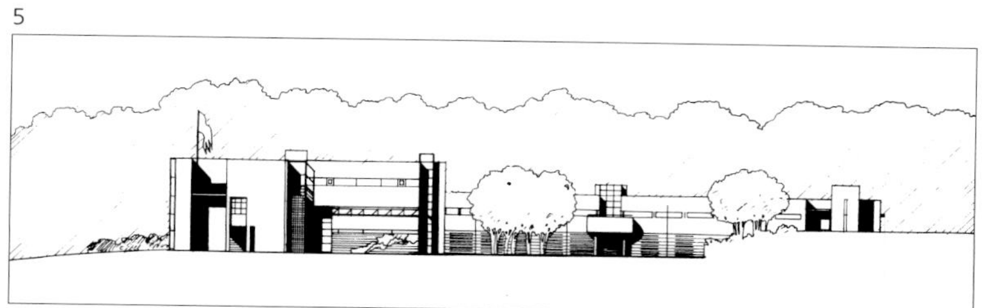

6

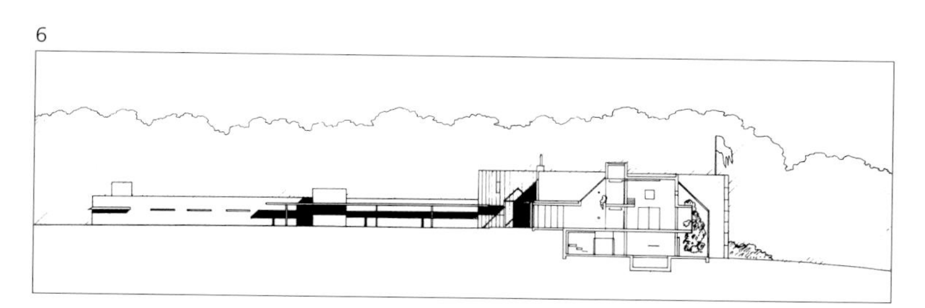

7

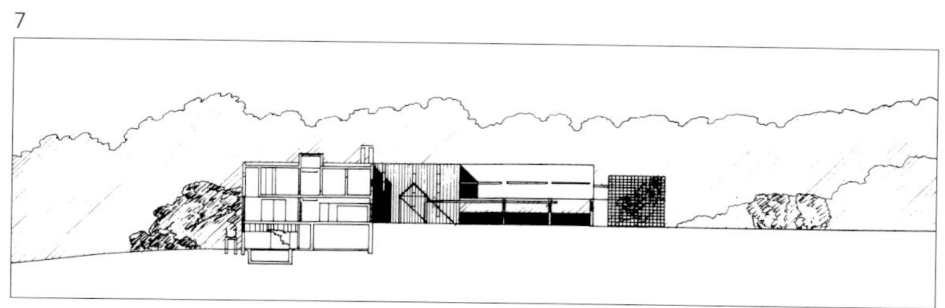

8

10

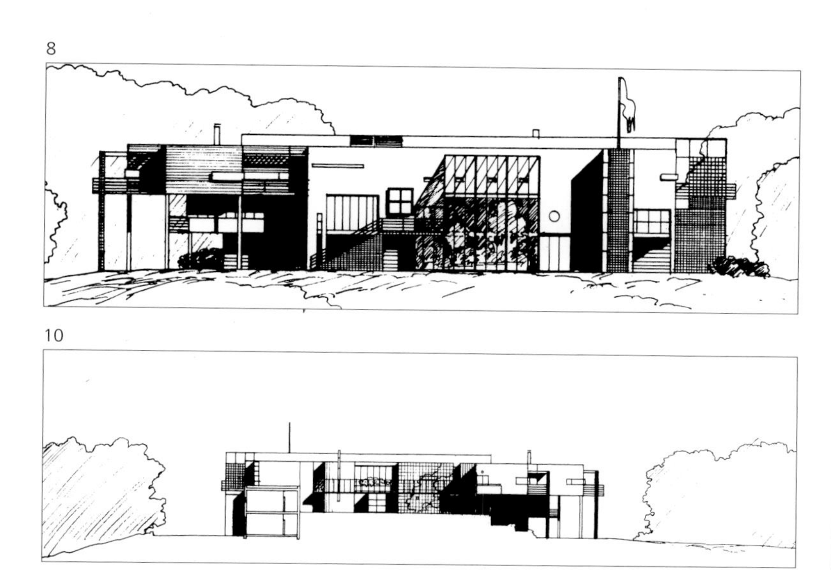

9

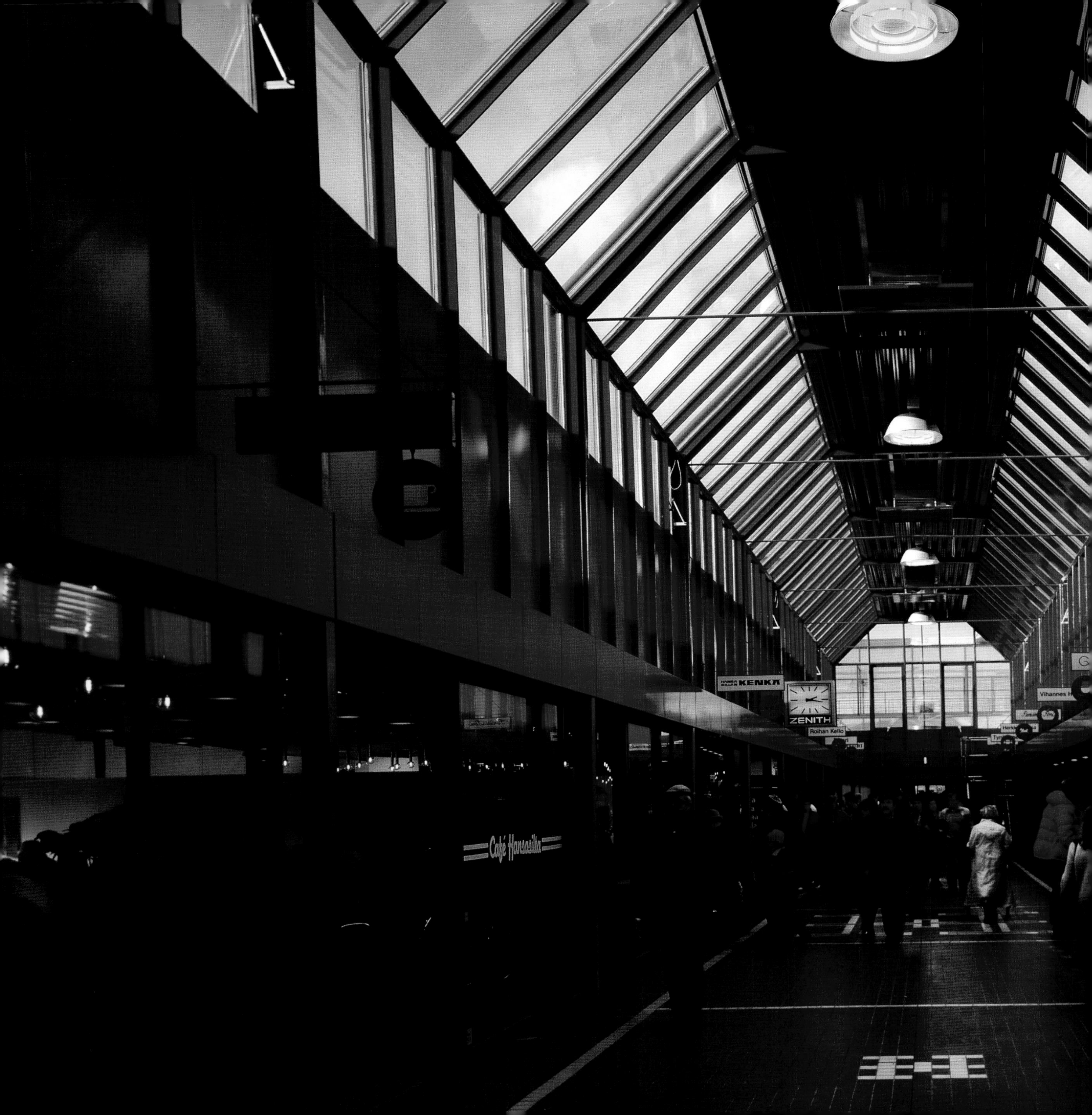
KENKA
ZENITH
Roihan Kello
Café Hansasilta

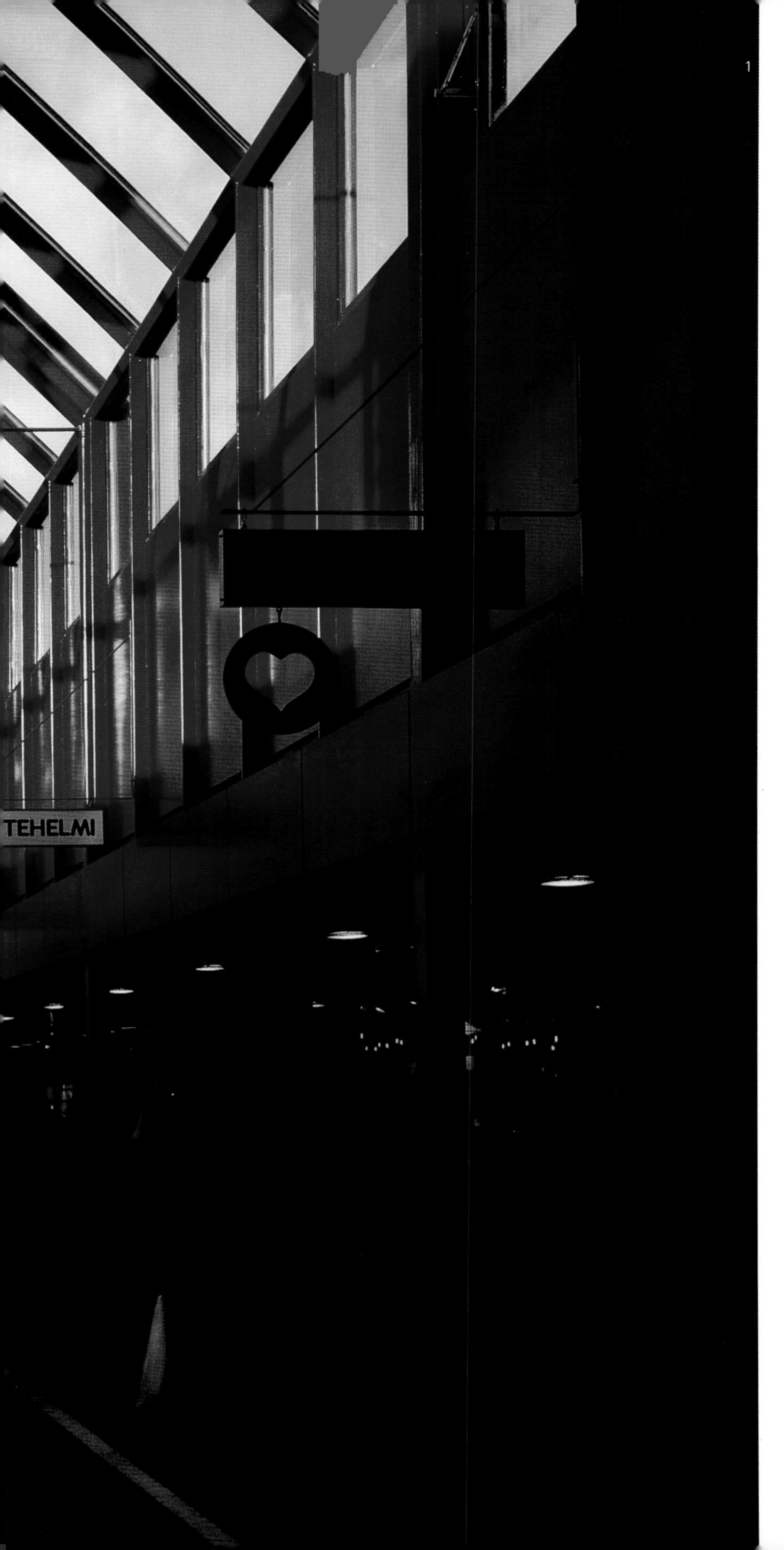

1

汉萨西尔塔，步行桥和购物商场，赫尔辛基 1984

Hansasilta, Pedestrian Bridge and Shopping Mall, Helsinki 1984

Itäkeskus购物商场已经发展成赫尔辛基东部最重要的中心之一。该中心提供大量社区服务，并产生了8000个新的工作职位，其中大多数职位在办公室和小型工业。在该商业服务圈辐射范围内的Itäkeskus的住宅区大约居住有150000人。

汉萨西尔塔桥位于最初的两个商业区之间，是一条重要步行路线的组成部分。在桥的设计中，共规划了13个单位，每个单位的平均面积为60m²。最初的想法是把桥设计成一个购物商场，一条由各种小商店组成的、有顶的步行街。商店就直接开在步行街上。桥的主要任务是划分步行区和机动交通区，这就意味着驾车者和行人对桥有着完全不同的视觉感受。对于驾车者来说，它就像一个桥式大门、一个进入赫尔辛基市的地标和象征，他对桥的认知来自桥长方向的立面和低部位的表面肌理。对于步行者来说，桥则是Itäkeskus 步行中心路线上的一部分、是一条按步行者尺度建造的布满商铺的街道，它将机动交通和噪声隔绝在外。步行者通过咖啡馆等节点位置的开口得以确定自己在环境中所处的位置。

建筑由两个主要部分组成：巨大的混凝土桥面，和生动有趣、色彩艳丽的金属覆盖结构。

所有的窗框以及从内部看得见的金属面板和钢支撑都刷成红色。步行街内部就像一种“温暖的外部空间”，它所有的结构和细节处理看起来都像一个开放的街市，但又不受气候条件的制约。建筑的外部是氧化铝面板。

Itäkeskus Shopping Mall has been developed into one of the most important centers of eastern Helsinki. The center provides numerous public services together with about 8,000 new jobs, mostly in offices and small-scale industry. In the new residential blocks in Itäkeskus live around 150,000 people within the radius of the commercial services.

Hansasilta Bridge is situated between the first two business blocks and is part of the most innportant pedestrian axis. In the design of the bridge, space was provided for 13 business premises, each with an average of 60m². The basic idea behind the bridge was a type of market hall, with many kinds of small shops gathered along a covered pedestrian street. The shops open directly onto the pedestrian axis. The main task of the bridge is to separate the pedestrian and motor traffic areas, meaning that the driver and the pedestrian each gets an entirely different view of the bridge. To the driver it is a bridge-like gate,a kind of horizontal landmark and symbol of arrival into the city of Helsinki. He experiences and recognizes the building through its long elevations and its lower surface 'elevation'. To the pedestrian the building is part of the central pedestrian axis of Itäkeskus, a covered shopping street, screened from the noise of the traffic and built to a pedestrian scale. The pedestrian defines his position in relation to the external environment and the roads only through the views opening up from the joints of the building and the cafeteria.

The building consists of two main parts: the massive concrete deck of the bridge and the playfully vivid and colourful metal clad structure around the steel deck.

All the window frames and cladded panels visible from the inside and joining the supporting steel structures have been painted red. The inner pedestrian street is a kind of 'warm outer space', which both structurally and in all its details is like an open street market, but without the limitations imposed by the climate. The building's exterior metal cladding consists of brown anodized aluminium panels and frame structures.

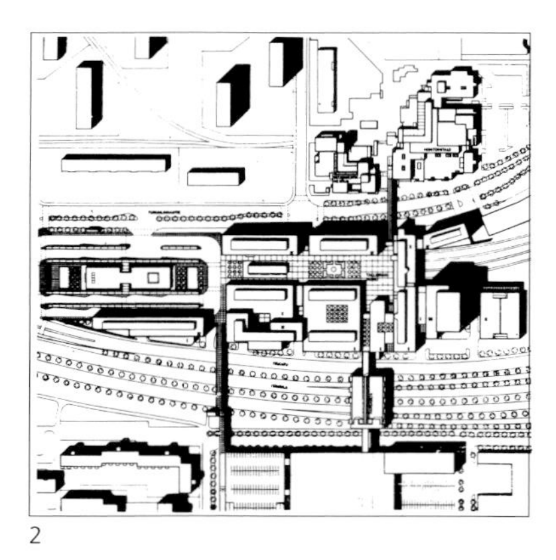

2

1 带商店的步行街
2 总平面
3 模型照片
4 平面
5 外观细部
6 剖面
7 剖面
8 购物中心
9 透视

1 Pedestrian street with shops
2 Site plan
3 Model photo
4 Floor plan
5 Exterior detail
6 Section
7 Section
8 Shopping mall
9 Perspective

3

4

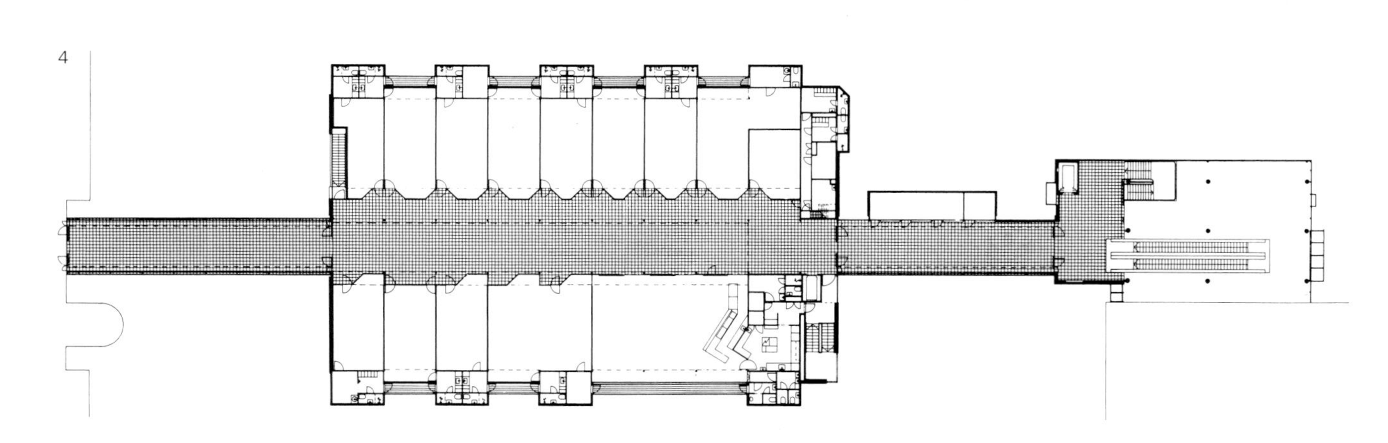

5

6

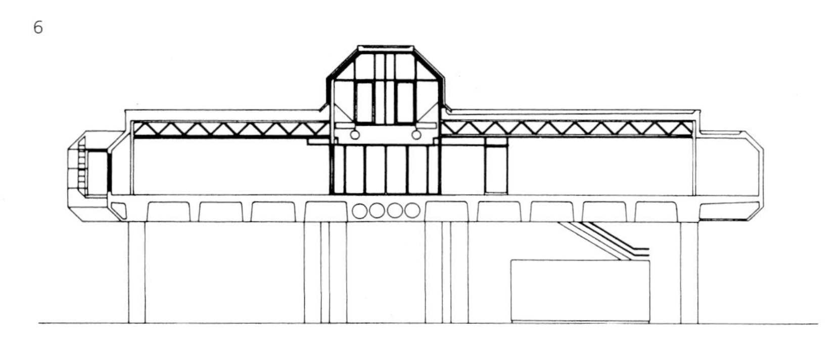

7

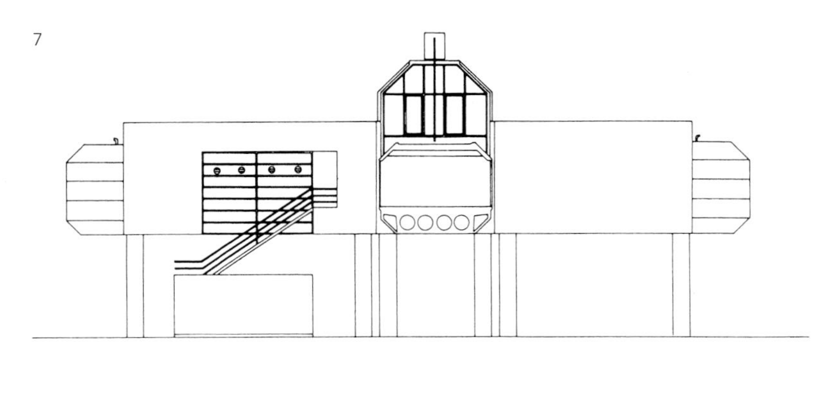

8

9

10

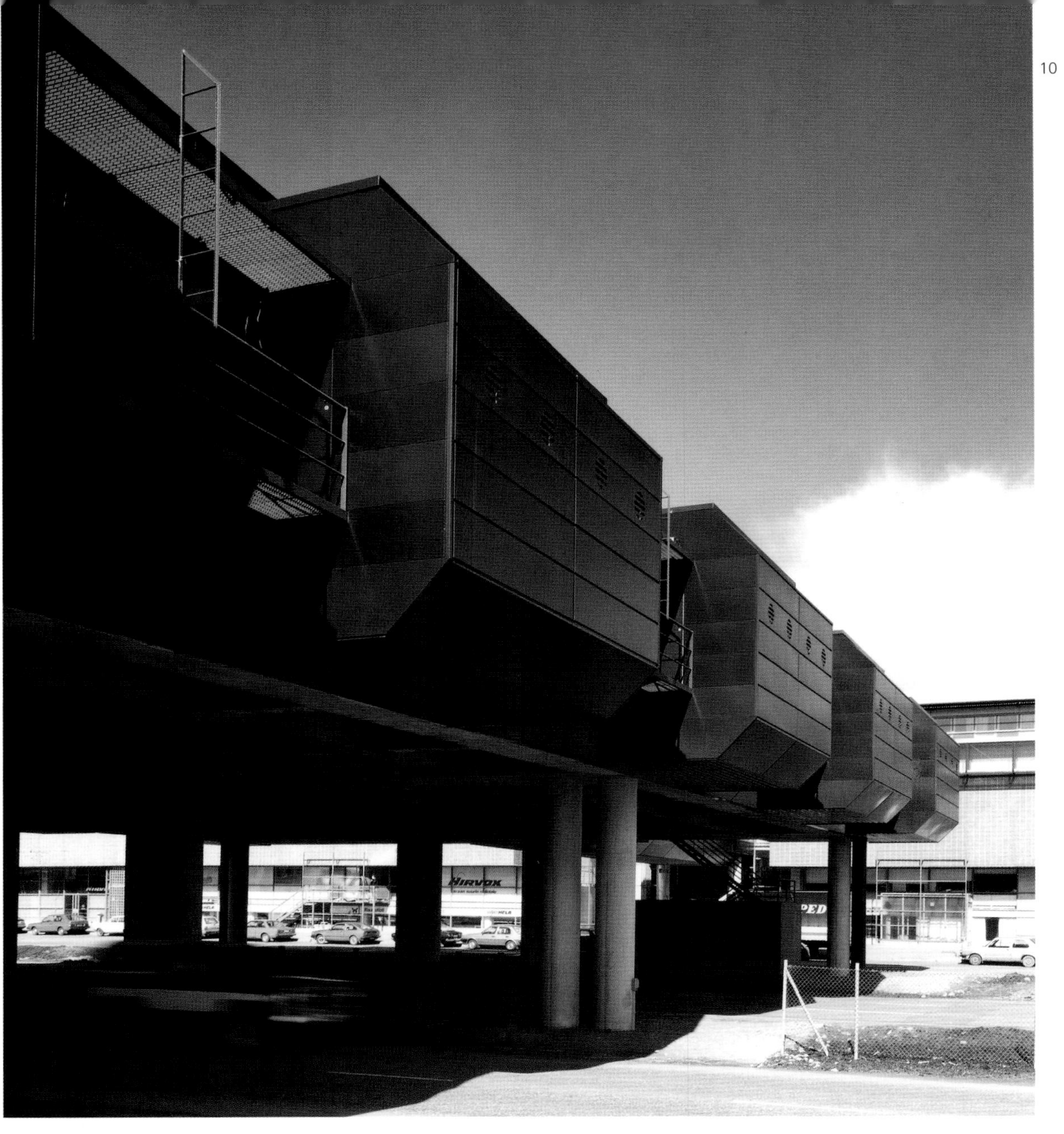

11

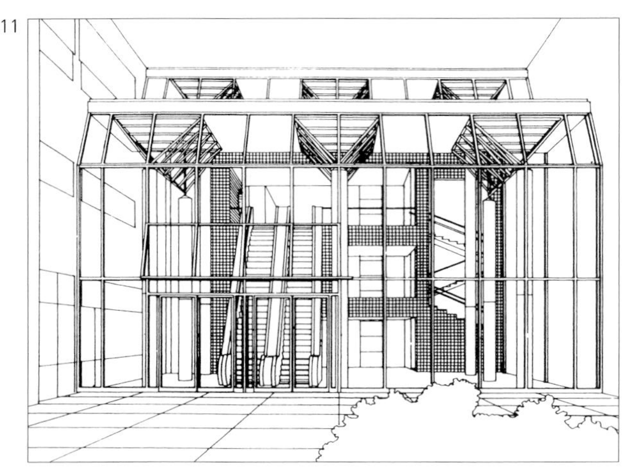

12

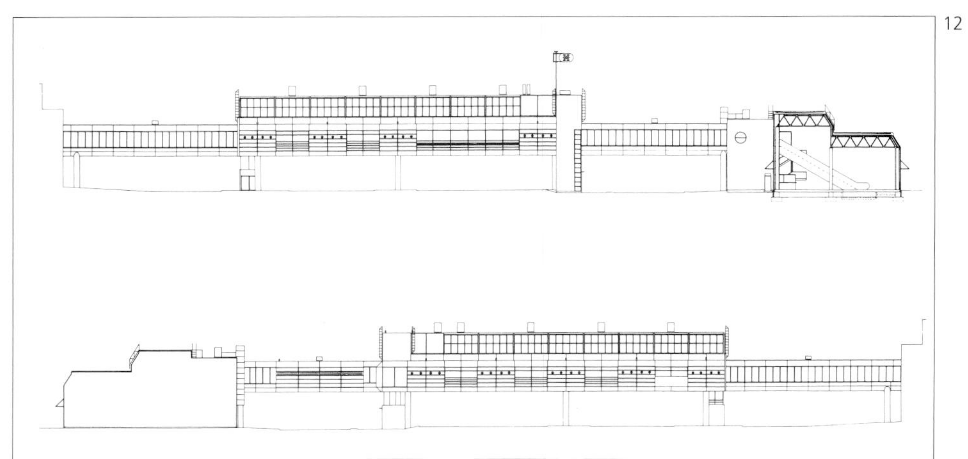

10 外观
11 主入口大厅
12 立面
13 入口大厅
14 透视
15 咖啡吧
16 入口大厅

10 Exterior view
11 Main entrance hall
12 Facades
13 Entrance hall
14 Perspective
15 Cafe
16 Entrance hall

13

14

15

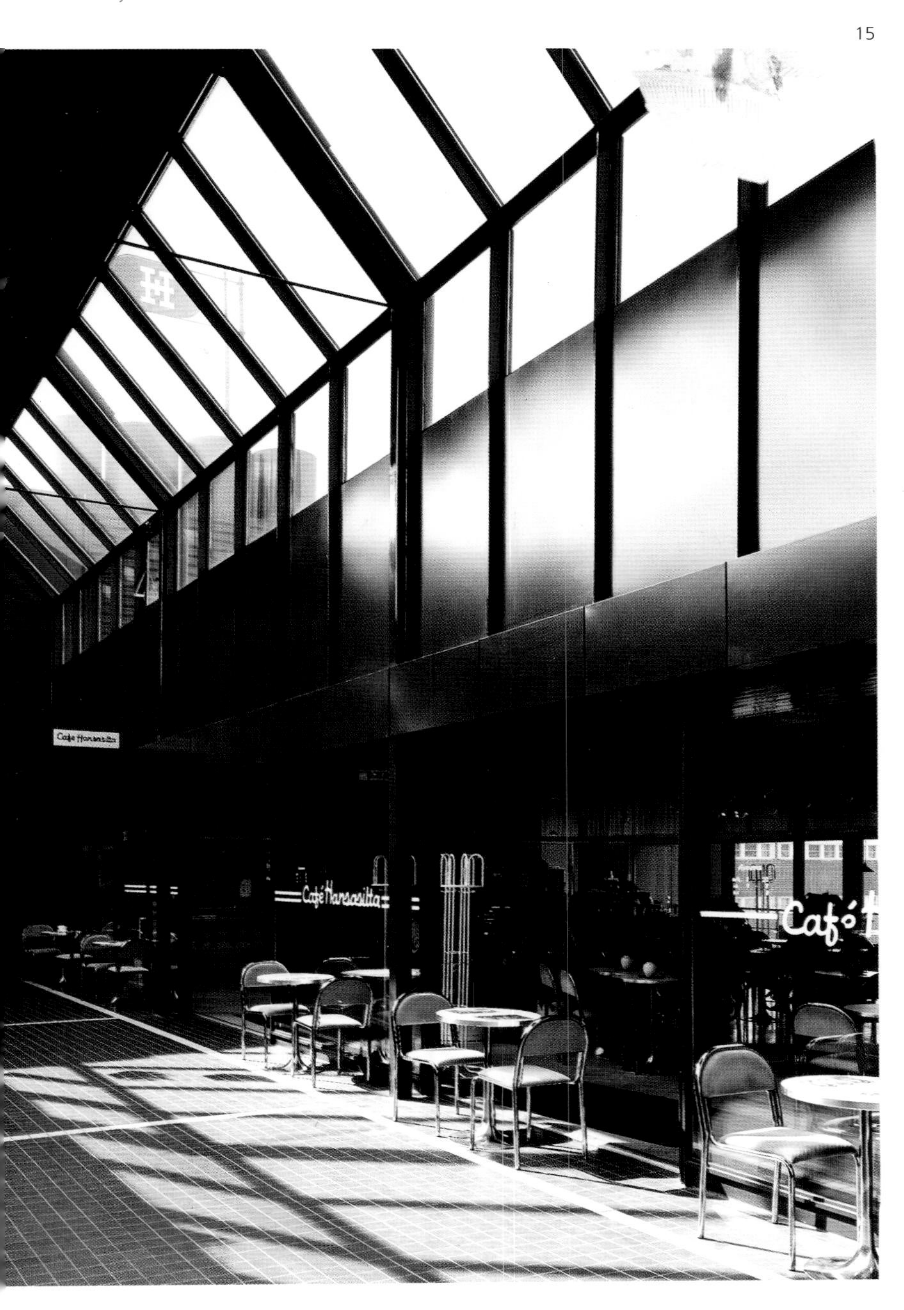

16

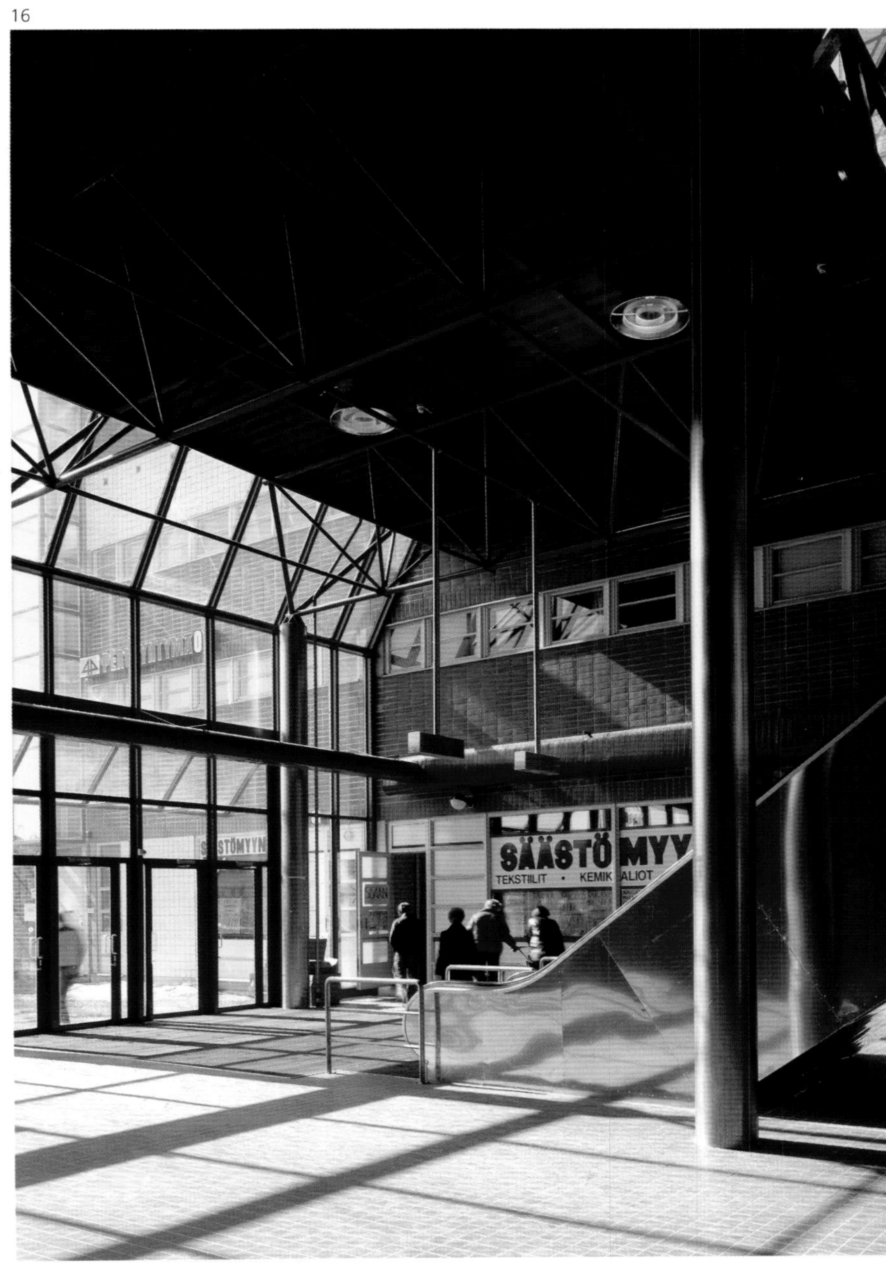

拉亚萨洛教堂，设计竞赛第一名，赫尔辛基 1984

Laajasalo Church, Competition entry 1st prize, Helsinki 1984

这个建筑位于一处需要尽可能多地保留绿树的场地上。而这个树木繁茂的地方正是建造高质量教堂花园的理想场所。

地形决定了建筑的主要入口及前院位置。

这个建筑由一个教堂和一个俱乐部会所组成，两者之间由休息大厅相连。

教堂本身有不同尺寸的三个厅。目的是使得每一个位置都可以很好地获得置身教堂的感觉，并确保座位与祷告台和唱诗班／风琴区域之间的视觉联系。

教堂的顶棚是白色弯曲的木格栅，混合了一些偏蓝的色调。圣坛的墙是由梁和砖墙在不同层次上交叠构成的。圣坛后的墙与一股流淌着的水体相结合：水从位于高窗之下的水源处流下，在墙脚的地方，水流、柔和的小瀑布和植物融合在一起。

每一个厅都面对着圣坛的墙，正是墙的各种细节给了各个厅不同的特性。

建筑主体结构是混凝土。钟塔局部是钢结构。教堂的立面和内表面材料是白色的砖和艳丽多彩的陶瓷。

The building is situated on the plot with the idea to conserve as many trees as possible. Consequently, the wooded area can be used as a basis for a high-quality citylike church park.

Functionally the site of the building determines the maintenance and entry side of the plot with leisure and entrance yards.

The building is divided into a church and clubhouse with a foyer connecting them.

The church itself consists of three halls of varying sizes. The aim was to create a sense of church space to all seats and to ensure good visual contact with the altar and choir/organ areas.

The ceiling of the church is a blend of white curved wood grilles and a bluish colour. The altar wall is built multidimensionally out of beam and brick wall elements placed at various levels. The wall behind the altar is bound by a freeform wall of running water: the lower part of the wall incorporates a stream of water with subdued waterfall and plant motifs, starting from a font below the tall window.

The halls all face the altar wall and it is the varying details of the wall that give each hall its own individual character.

The supporting structures are made of concrete. The campanile is partially steel-structured. The materials of the elevations and the church interior are white brick and pastel-coloured ceramic tiles.

1

1 外观透视
2 模型照片

1 Exterior perspective
2 Model photo

3 总平面
4 平面
5 室内透视

3 Site plan
4 Floor plan
5 Interior perspective

3

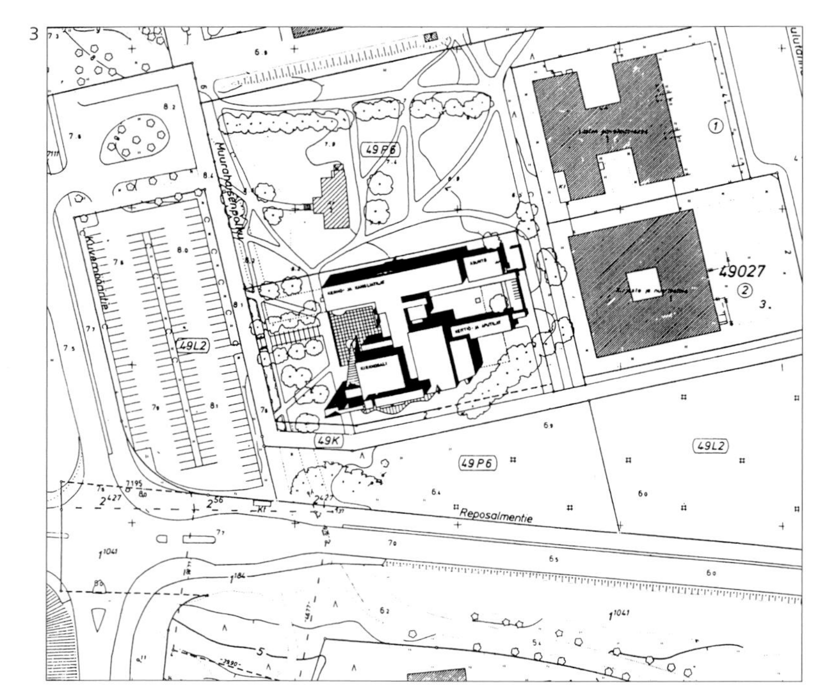

5

4

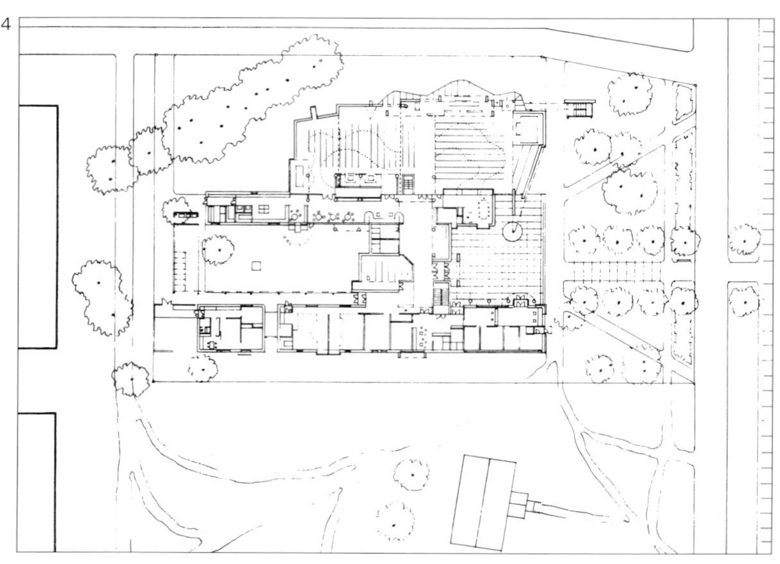

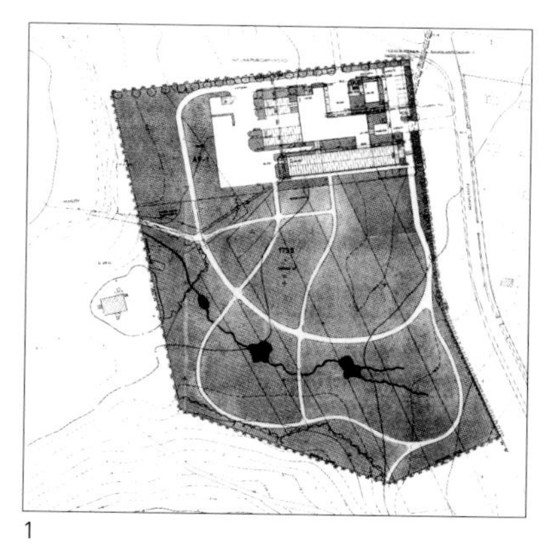

1

2

3

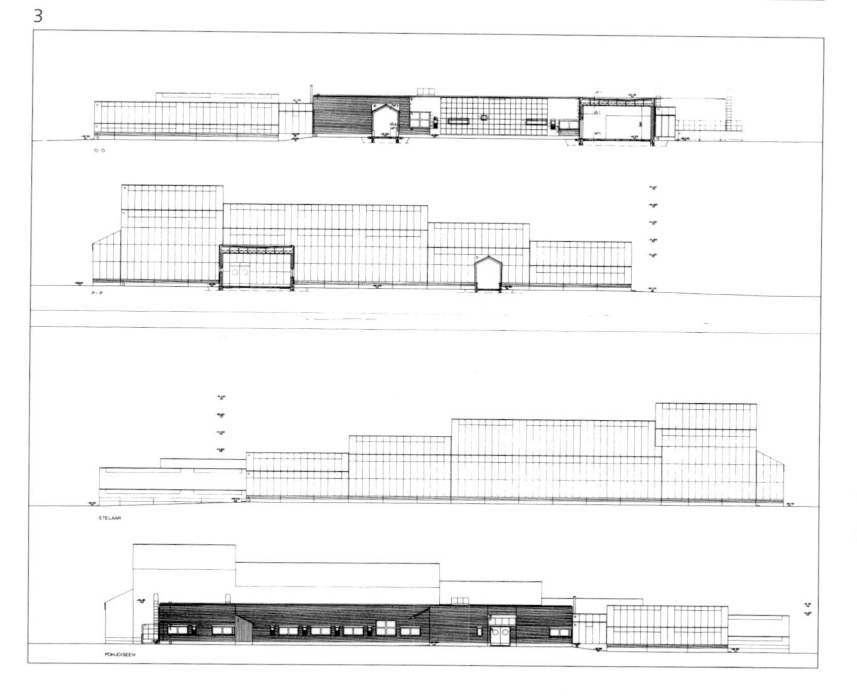

植物园和温室，约恩苏大学 1985

Botanical Garden and Glass House, Joensuu University 1985

1 总平面
2 平面
3 立面

1 Site plan
2 Floor plan
3 Facades

约恩苏植物园位于Lintulahti地区，靠近约恩苏漫步道和露天唱诗台。它既能满足大众的休闲和文化需求，也可以成为科学研究基地。

温室的作用能够满足人们的好奇心和研究兴趣，也为不同的植物创造了生存所需的特殊环境。通过科技手段模拟自然环境，使得原来分布在世界各地的各种植物汇聚在一个小空间里供研究和展览。一般建筑中忽略的东西，即自然的植物群落，在设计中上升为最主要的建筑元素。所有对公众开放的部分都设计成适合植物生长的环境。另一方面，设计也利用和发挥了砖结构、平屋面的服务用房，和玻璃结构、坡屋面的温室两者在外形上的对比。

基于这些设计原则，建筑的正面是棚架结构，主入口的门廊也是温室的形式，在展览建筑和服务建筑之间形成一个庭院，庭院部分的建筑外表面运用了白色的砖，这样经过白色表面漫射出来的光使得这个庭院比其他部分的室外空间更温暖一些。

功能上，建筑结合了两个重要因素：一方面是研究和公共人流对空间和空间关系的要求；另一方面是植物对生长环境和自然光照的要求，以及相关的一套复杂的机械、维修和控制设备。

温度最高、湿度最大、最具热带风情的展览部分位于建筑的中部和入口处。干燥而温度较低的展区位于后部。

Joensuu botanical garden is located in the Lintulahti area, near Joensuu trotting track and the open air choir stage. Consequently, it caters for the leisure and cultural needs of the public as well as scientific research.

The glass house is a unit formed to serve human curiosity and research and the special environments required by the plants. A broad and varied selection of plants for research and exhibition is assembled within a small space, using natural conditions aided by technology. The dimension missing from a conventional building – the world of flora – has been elevated to the status of a central architectural element in the basic design. All the structures and surfaces open to the public have been designed to serve the growth environment of the plants. On the other hand, the design exploits the contrast between the flat-roofed brick maintenance buildings and the pitched-roof glass houses froming the external image.

Following these principles, the façade of the building is covered with pergola structures, the vestibule of the main entrance is built like a glass house, and the courtyard between the exhibition glass house and the maintenance building has been turned into an area warmer than the other outside spaces by using a white lightdiffusing brick facing.

Functionally the building design is formed by coordinating two elements: The space and interconnections required by research and public traffic on one hand, and the natural lighting and growth environments of the plants and the complicated technical equipment and maintenance and control systems on the other.

The warmest, most humid and most exotic tropical department with its showcases is situated in the entrance yard and the foyer. The drier and cooler departments come logically after the tropical area.

4 玻璃房夜景
5 从南部看
6 玻璃顶棚
7 热带花园
8 从花园外部看
9 面向东部的景观
10 远望湖面

4 Glass house at night
5 View from the south
6 Glass ceiling
7 Tropical garden
8 View from the garden outside
9 View towards the east
10 View to the lake

5

6

7

8

9

10

森林研究中心，坎努斯 1985

Forest Research Center, Kannus 1985

坎努斯研究中心的主要任务包括一般性的地方森林研究，和针对Ostrobothnian滨海地区的特殊研究两大类。因为对森林能源的研究正在进行中，因此中心至少在开始阶段会注重木材作为一种能源的研究。

研究中心的设计以"图式语言"为指南，从人类对舒适和实用的基本要求出发，制订切合森林研究中心特殊性的、目标明确的设计方案。

研究站包括以下功能：办公室、实验室、车库、操作室与储藏室、温室、以及工作人员宿舍。

根据Ostrobothnian地方特点，研究中心采用了较封闭的格局。办公主楼两层高。靠近主楼的温室和实验室为一层。红砖、赭石色的防护板和黑色金属屋面构成了建筑的主要色调。

主楼的室内为白色，格栅的形式符合建筑关于木材的主题。

The main task of Kannus Research Center comprises regional and provincial forestry research projects in general and the problems of the Ostrobothnian coastal area in particular. Because of the forest energy research programme currently in progress, the center will, at least in the beginning, concentrate on research into wood as a source of energy.

To allow more through goal-oriented planning design goals peculiar to Forest Research Center were developed using 'pattern language', which means emphasizing the important human aspects of comfort and functionality.

The station has the following facilities: offices, laboratories, garages, workshops and storage facilities, a glass house and the caretaker's apartment.

The buildings form a partially closed group following the Ostrobothnian tradition. The main building, containing offices, has two storeys. The laboratories and glass house adjacent to the main building, and the separate living quarters and maintenance building, all have one storey. The main colours of the station are red brick, red ochre coloroured weatherboarding and black metal roofs.

The interior of the main building is white, emphasized by the various grilles and lattices made, true to the building's use, of wood.

1

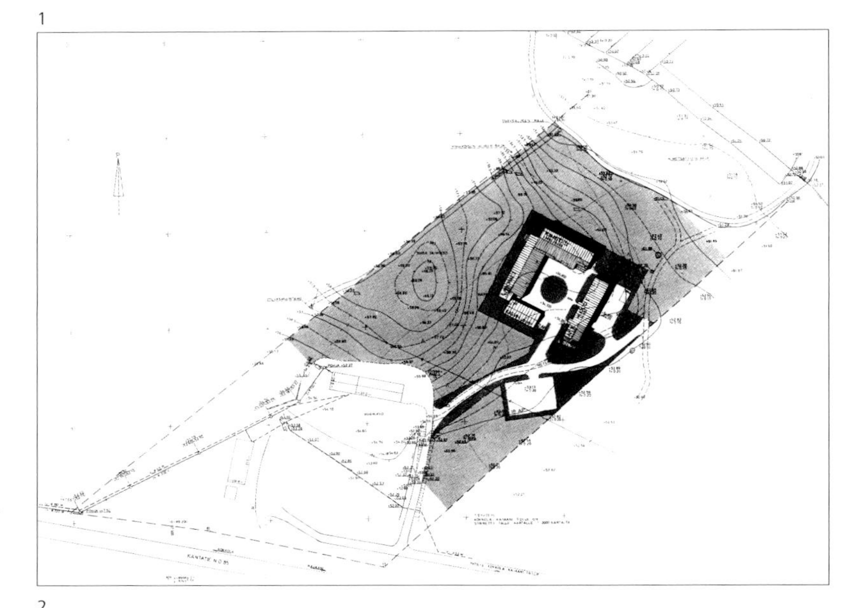

2

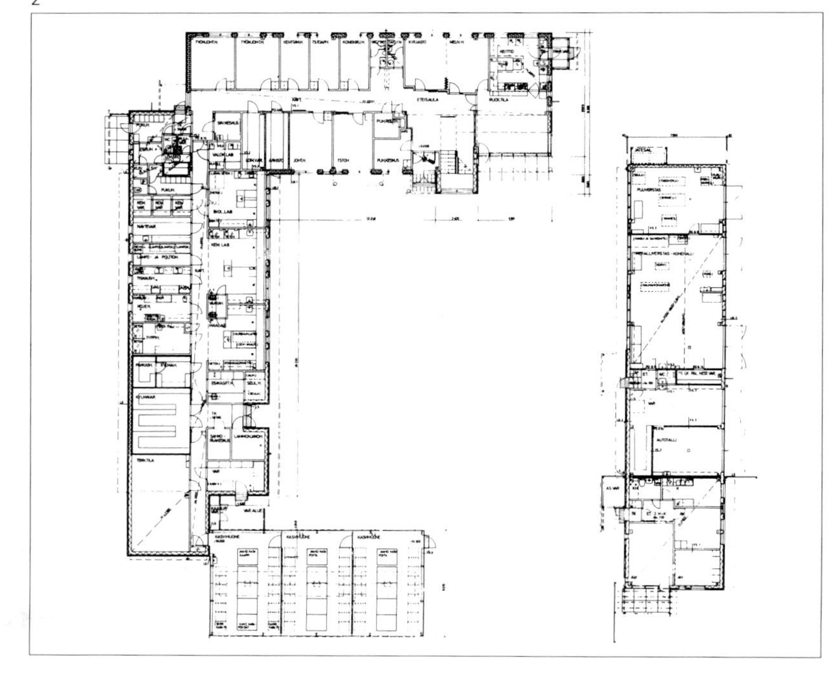

1 总平面
2 平面
3 外观

1 Site plan
2 Floor plan
3 Exterior view

4

5

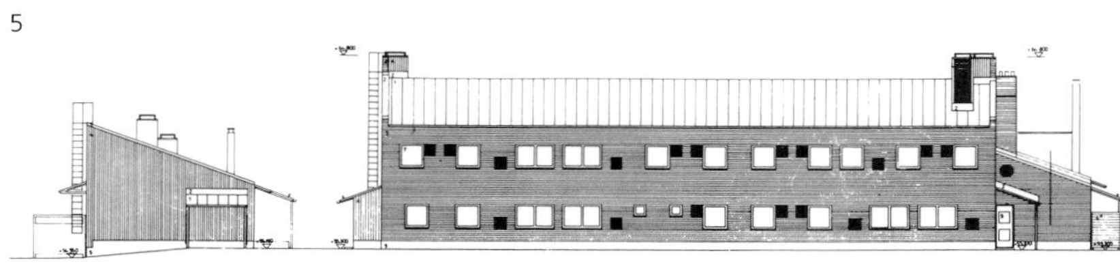
JULKISIVU KOILLISEEN

6

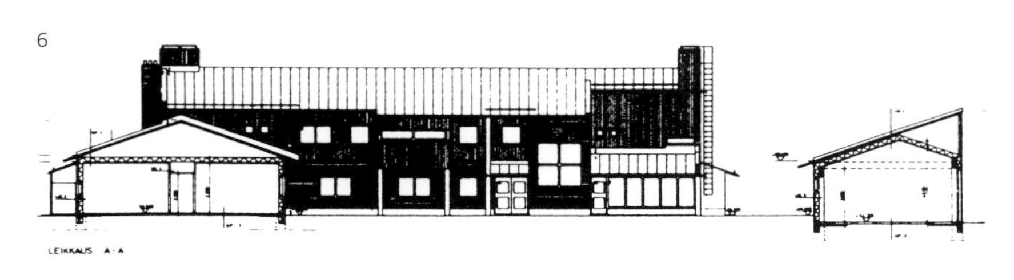
LEIKKAUS A-A

7

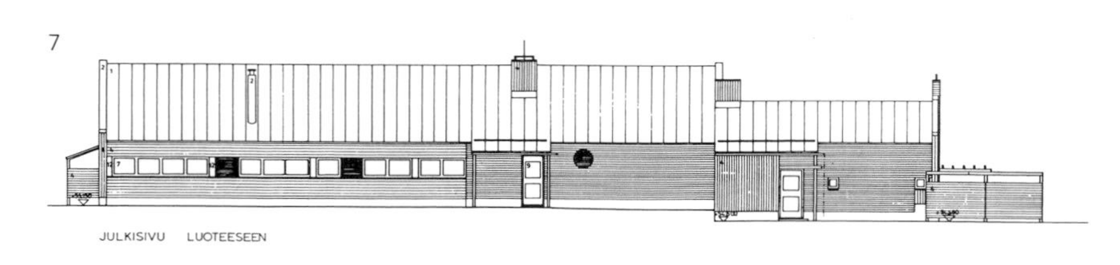
JULKISIVU LUOTEESEEN

8

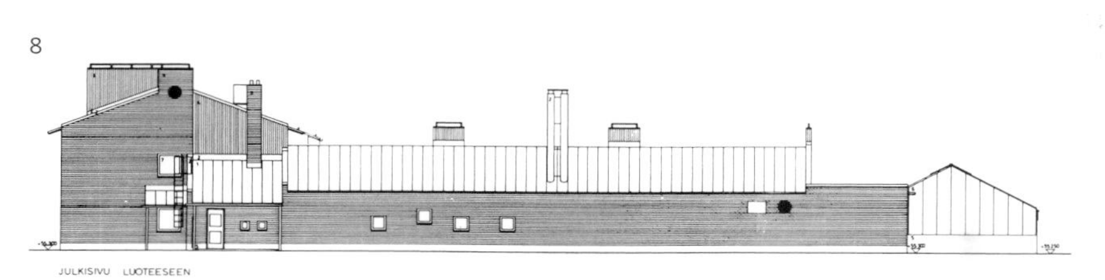
JULKISIVU LUOTEESEEN

9

4 室内透视
5 东北立面
6 剖面
7 西北立面
8 西南立面
9 夜景
10 从西南处看

4 Interior perspective
5 North-east facade
6 Section
7 North-west facade
8 South-west facade
9 View at night
10 View from the south-west

10

科伊维科私人住宅，赫尔辛基 1985

Private House Koivikko, Helsinki 1985

这些在第二次世界大战后建于赫尔辛基Pakila的"一层半"的住宅已经成为这个地区的标志物。从20世纪60年代初开始，在规划的纵容下，建设行为进入了一个混乱、无计划的状态。科伊维科住宅表现的是一种改造的思路。它试图回答的问题包括如何使加建的房屋与周围环境相适应，同时又考虑影响建筑设计的多种因素，诸如传统、气候、视野和心理形象等等。

房子的内部是典型战后时期的布局风格，房间围绕着一个中心火炉。两层高的起居室成为视觉中心，为传统风格题材增添了新内容。

建筑朝街道的一面比较封闭，而西面透过一个日光室向庭院开放。

科伊维科私人住宅是为一个四口之家按照较宽裕的空间要求设计的，它为将来通过分隔形成数个独立的单元提供了可能性。建筑同时表现了精美的制作工艺：地窖由混凝土砌块建造，地面以上部分为木结构，外部结合了金属构件。

The 'one-and-a half-storey' houses built in Pakila, Helsinki after World War II gave the area its image. Since the beginning of the 1960s, town planning permitting, started a confused, unplanned construction phase. House Koivikko is an expression of respect to the redevelopers. It is also a late reply to the question how additional building can be planned to suit the environment while taking into account factors which influence architecture, such as tradition, climate, views and mental images.

The interior of the house is a kind of revival of post-war rebuilding scheme: a centre chimney surrounded by rooms. The two-storey living room as a visual centre space adds life to the old scheme.

The house is closed towards the street and opens up through a greenhouse into the garden facing west.

The design of House Koivikko is based on the extra space requirements of a family of four, working on the assumption that it can later be used as a separate, self-contained unit. The house is also a tribute to craftsmanship: the cellar is built of concrete blocks, the floors above ground of timber structures with exterior steel trimming.

1

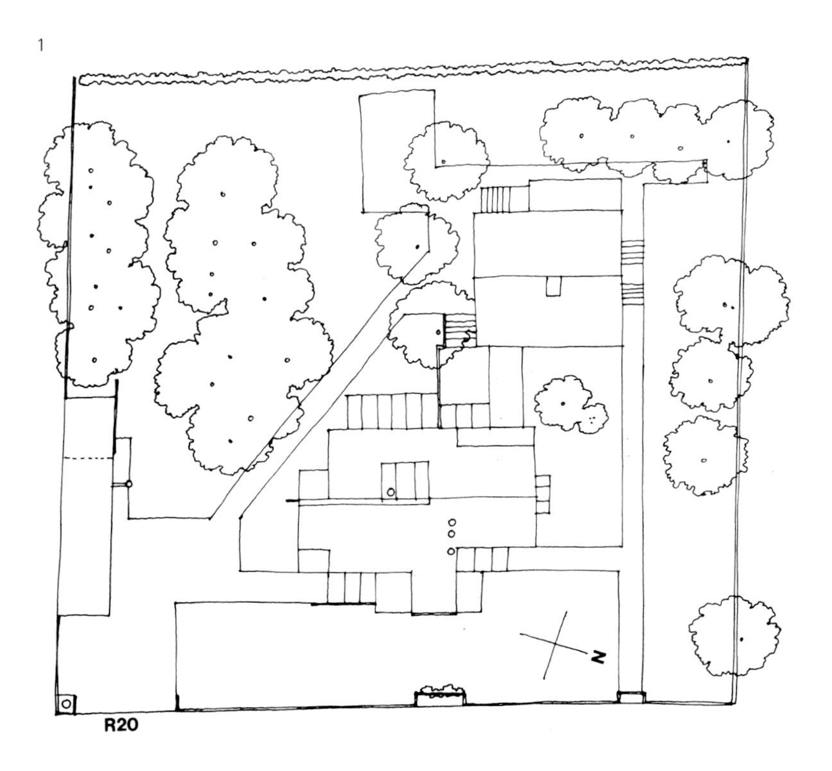

2

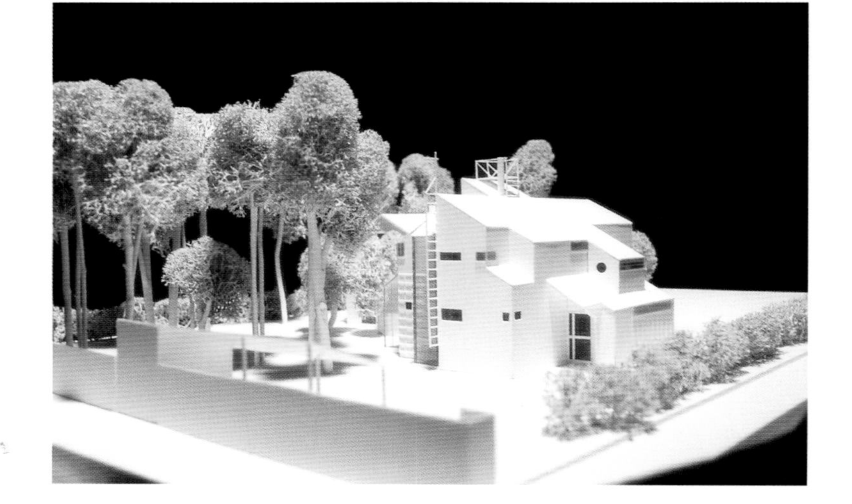

1 总平面
2 模型照片
3 温室

1 Site plan
2 Model photo
3 Greenhouse

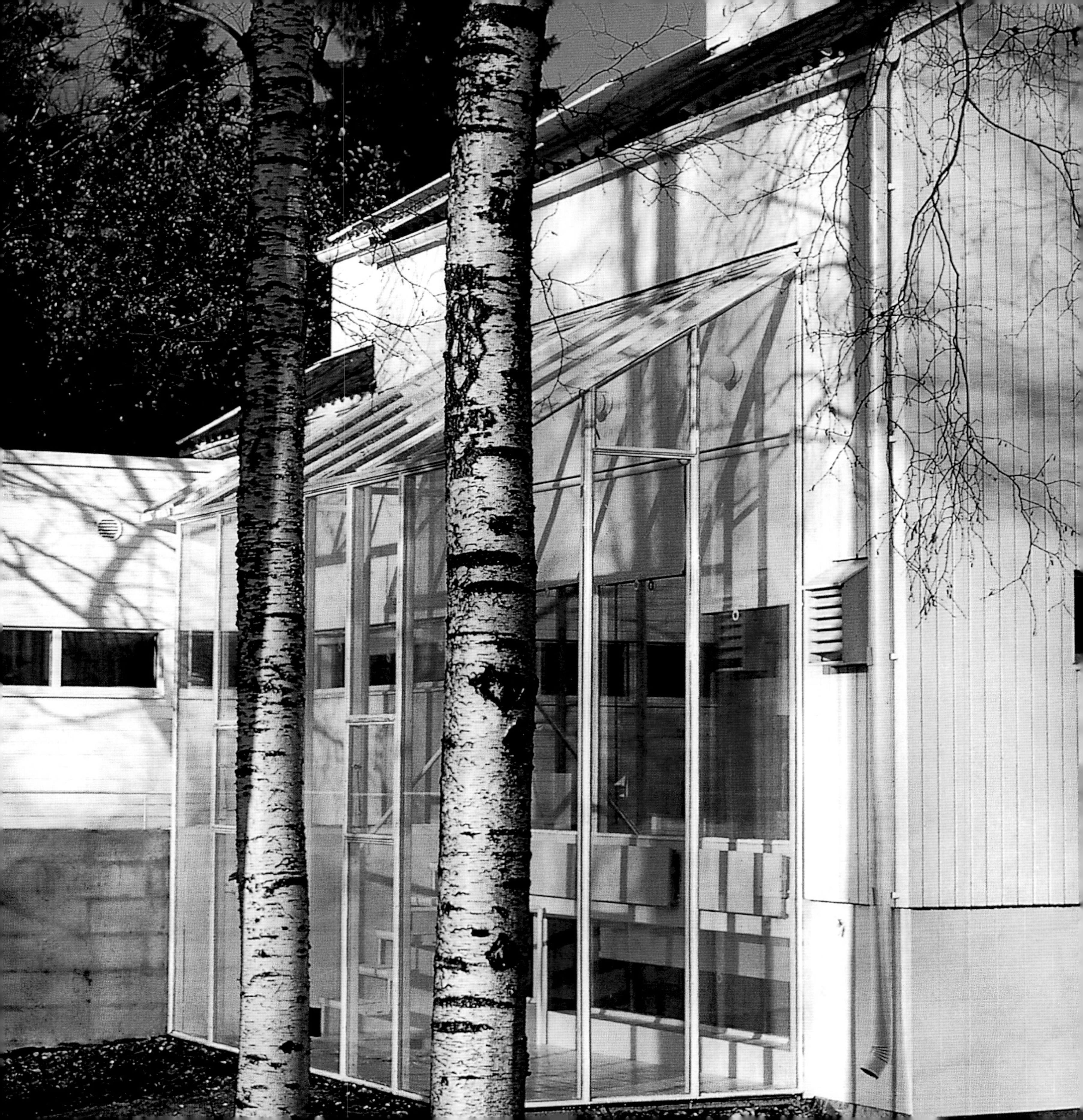

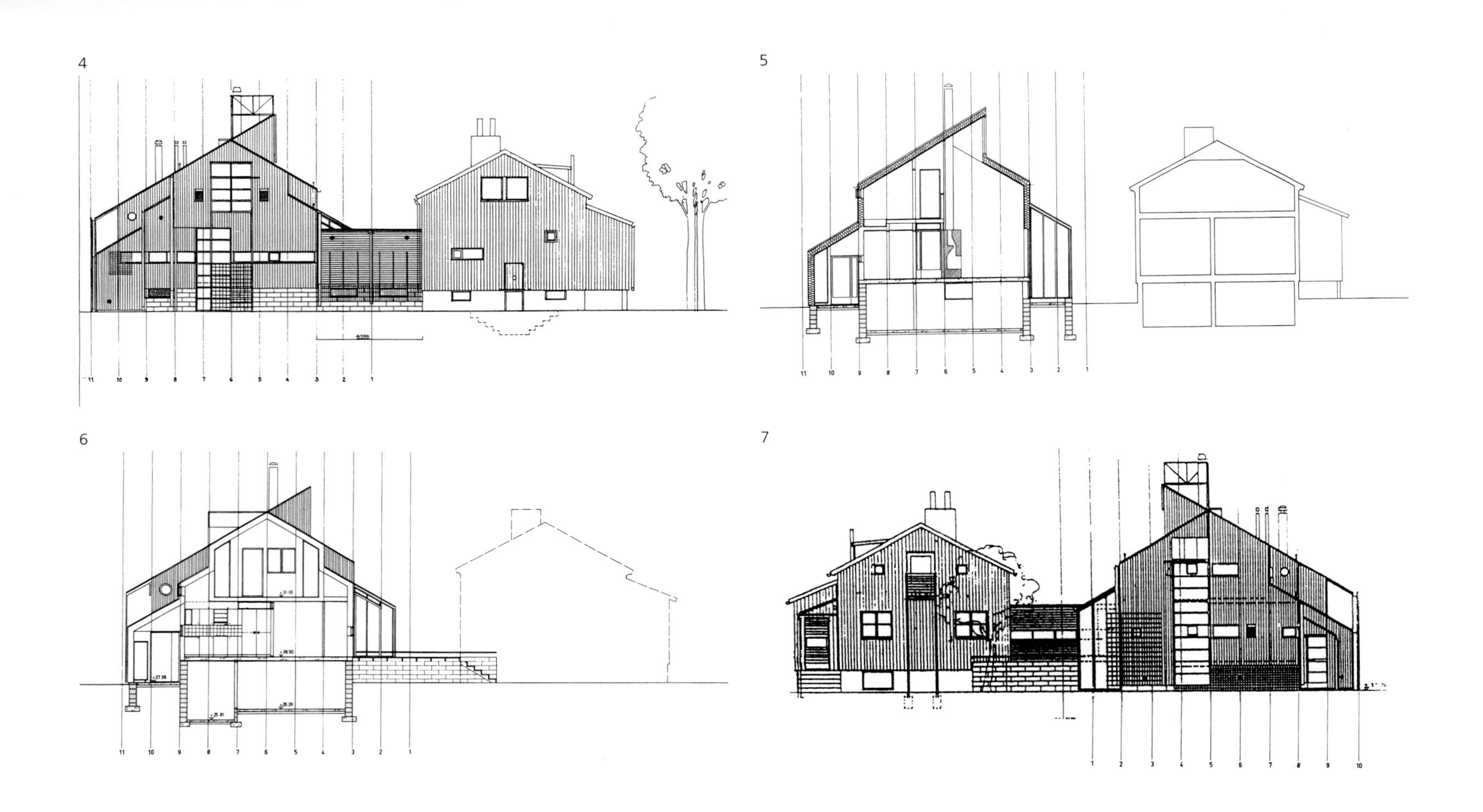
4
5
6
7
11
10
9
8
7
6
5
4
3
2
1

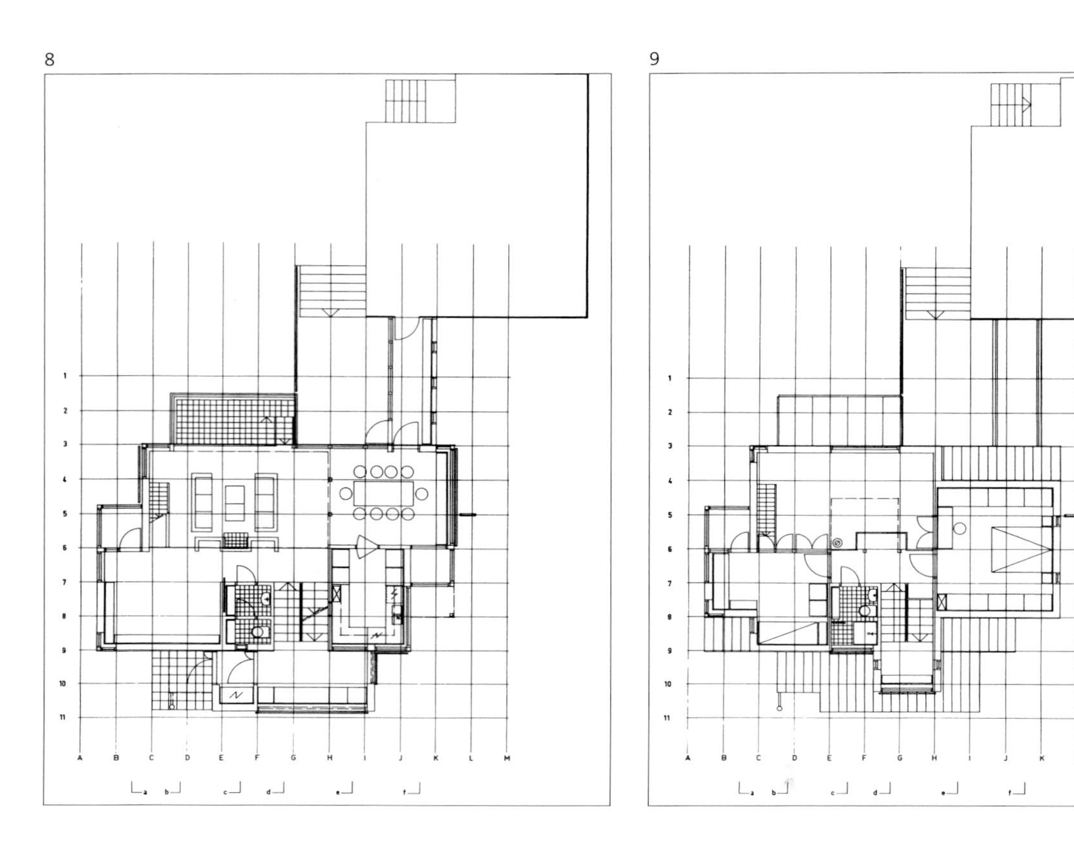
8
9
A
B
C
D
E
F
G
H
I
J
K
L
M
a
b
c
d
e
f

10

4 北立面
5 剖面
6 剖面
7 南立面
8 一层平面
9 二层平面
10 天窗
11 从街上看
12 庭院
13 从南面看
14 从内院看

4 North facade
5 Section
6 Section
7 South facade
8 First floor plan
9 Second floor plan
10 Skylight
11 View from the street
12 Courtyard
13 View from the south
14 View from the courtyard

11

12

13

14

15

16

15　从二层阳台看
16　起居室

15　View from the second floor balcony
16　Living room

库尔基马基多功能大厅，
赫尔辛基 1989

Kurkimäki Multipurpose Hall, Helsinki 1989

1 总平面
2 平面
3 主入口

1 Site plan
2 Floor plan
3 Main entrance

1

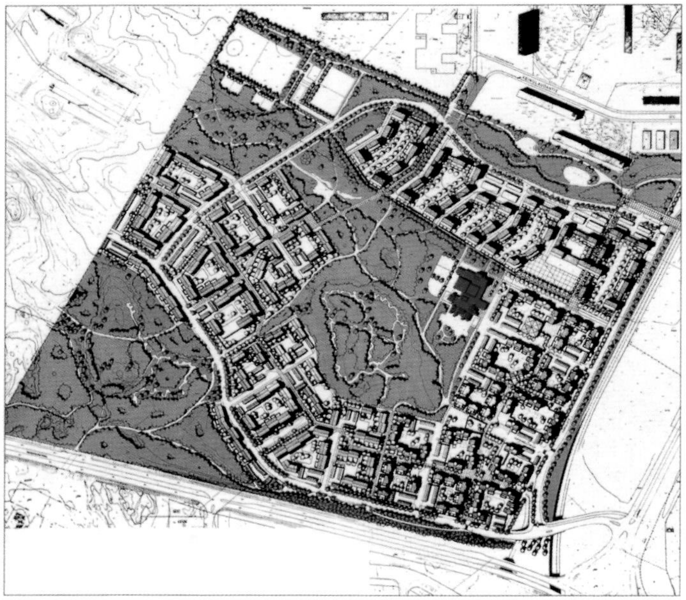

2

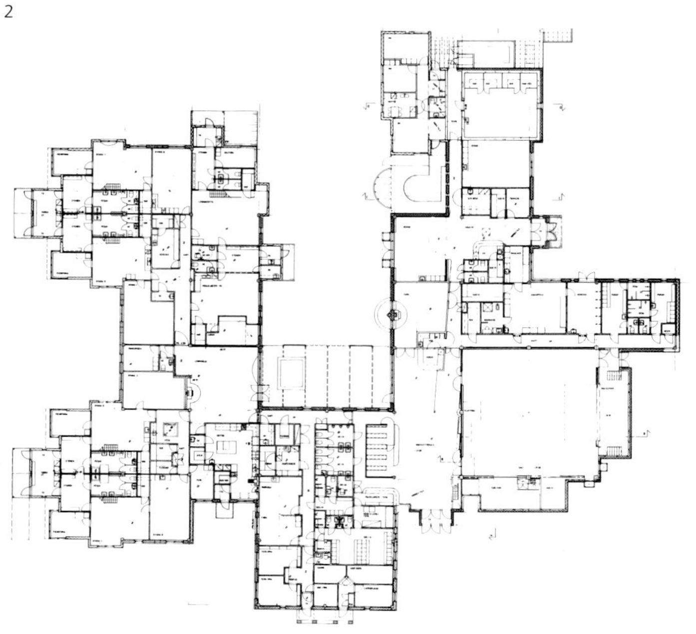

这是一个为小范围的居民提供服务和集会场所的社区中心。它包含多种功能以满足不同年龄层人士的需求，其中有幼儿日托中心、青年活动中心、集会厅和为成年人设立的业余活动室。建筑的目的就是让不同年龄和不同爱好的人们能够聚集在一起。因此，这个设计希望通过交流来发展和丰富各种活动并促进该地区不同年龄的人们之间的接触。

建筑的基本功能出发点一方面是建立各部分的个性，另一方面则强调建筑是一个整体。

设计以村庄的组织和视觉形象为隐喻。这需要某种尺度的转换：日托中心的每个单元成为村庄里的房子；走廊变为村庄里的道路；入口休息室就是村庄里的开放的公共场所；厅堂是村里一个有顶的集会场地，当然也可以被看作城堡中的庭院；风雨球场即长青公园；维修部就是村庄的工业大厦；储物棚则是户外凉亭。运动场是一个大院，售货亭则可以被看成是一个紧贴着公园的咖啡厅。

这种思路同样影响了细节的设计，并给日常生活添加了一点来自孩子们童话世界里的味道。

房子和大堂的各个部分由不同尺寸和颜色的砖块砌成。日托中心是由木头建成的，外面是柔和的浅色调，里面则是灰白色。维修部由金属板制成。

The building has been designed as a local service centre and meeting place for the inhabitants of a small housing area. It is composed of several functionally different elements, aimed at different age groups: children's day-care centre, youth facilities, assembly and hobby rooms for adults. The idea is to assemble different functions and age groups under the same roof. Consequently, it is hoped, the interaction will enrich various activities and promote contact between different age groups in the area.

The functional basis is to establish the identity and individuality of the various parts on one hand, and to create a concept of the building as a whole on the other.

The problem has been approached through the functional and visual conception of a fictional village. This leads to a sort of scale modification: the day-care centre units grow into buildings, the corridors become village roads, the entry foyer turns into the village common, the hall assumes the nature of a covered assembly place, a castle courtyard, if you like. The covered (bad-weather) playground becomes an evergreen park, the maintenance wing is the village's industrial hall, and the storage shelters are outdoor pavilions. The play hall is a yard, and the kiosk evolves into a market place cafeteria extending to the park.

This view has influenced the design of the details as well, and thus brings a bit of a child's fairy tale world into everyday life.

The basic part of the building and hall are made in different sizes and colours of brick. The day-care centre is built of wood, lightly toned on the exterior and bleached on the interior. The maintenance wing is of sheet metal.

4

5

6

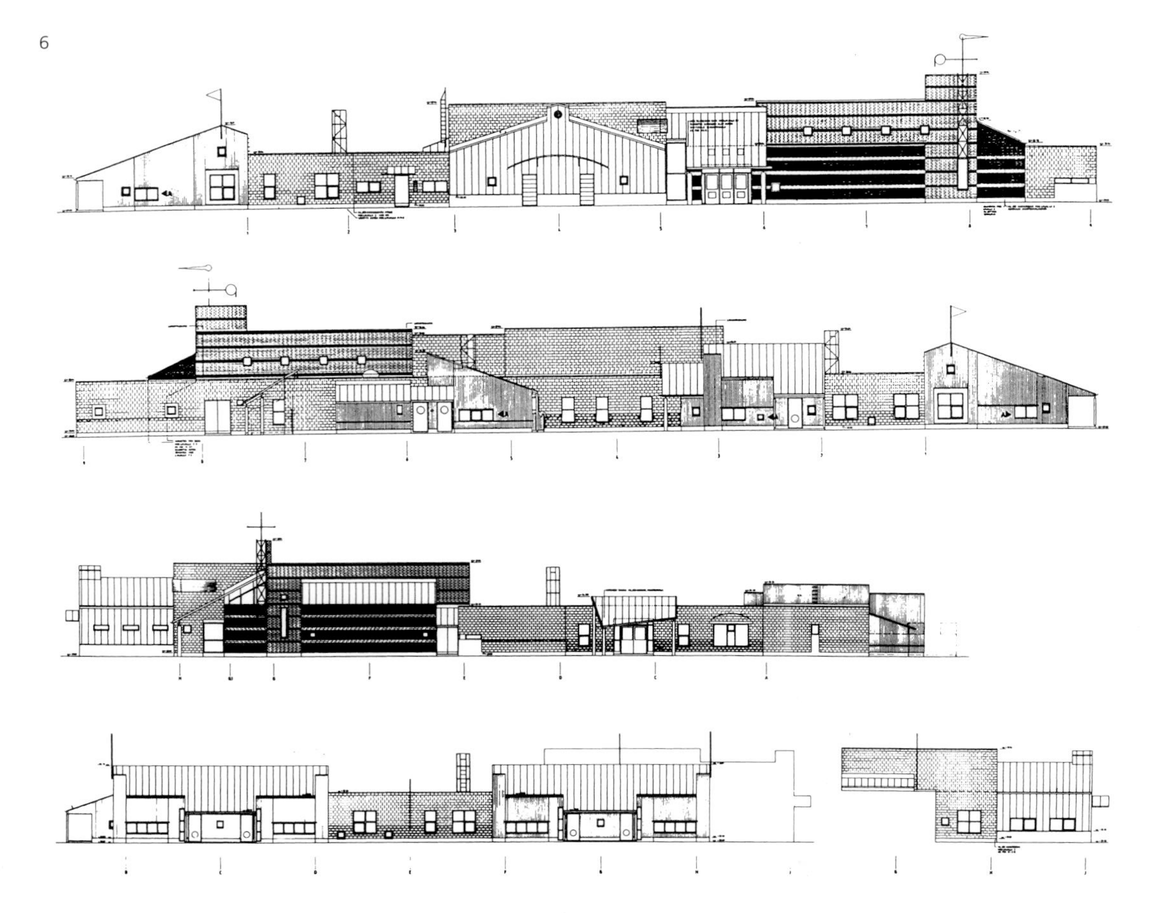

4 从东面看
5 日托中心立面
6 立面
7 透视
8 日托中心庭院

4 View from the east
5 Day care centre facade
6 Facades
7 Perspective
8 Day care centre courtyard

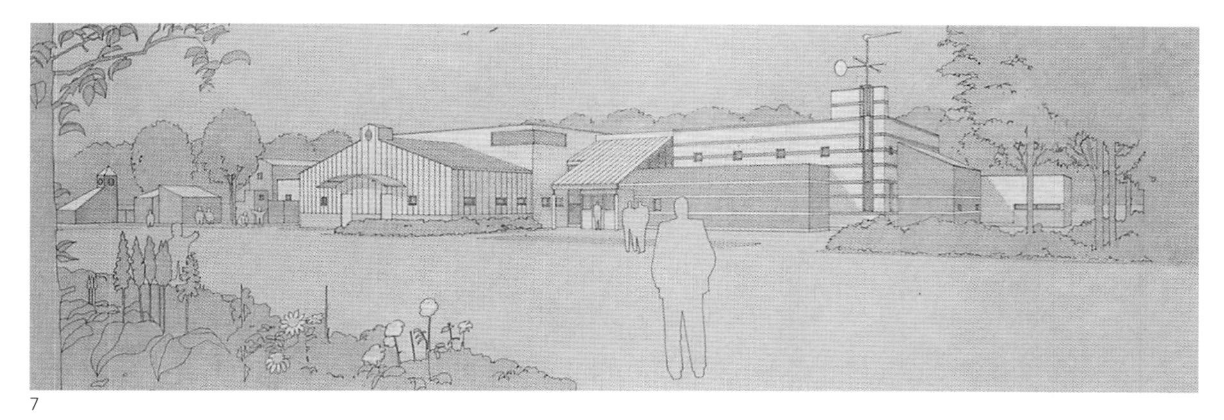

7

8

9　室内透视
10　主门厅
11　儿童餐室

9　Interior perspective
10　Main lobby
11　Children's dining room

9

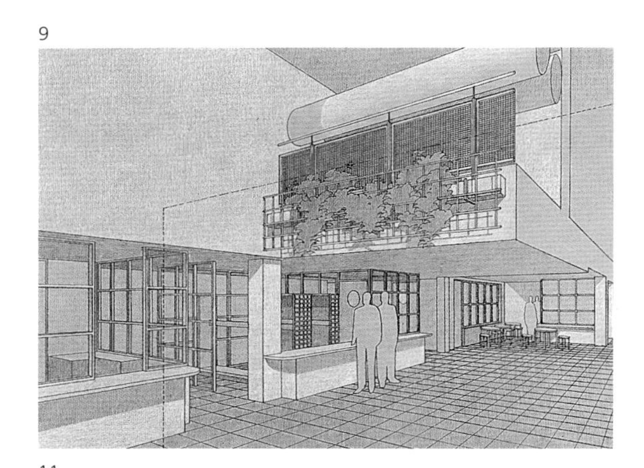

11

10

坦佩雷大厅，坦佩雷 1990

Tampere Hall, Tampere 1990

1 总平面
2 主入口
3 从公园内看
4 从街上看

1 Site plan
2 Main entrance
3 View from the park
4 View from the street

在1983年举办的全国建筑设计竞赛中，我们命名为“谢林工厂”的方案赢得胜利。

这一命名将该建筑与坦佩雷工业区联系起来。

由于选址在一块绿地上，因此可设计成完全独立的建筑。尽管体量巨大，但该建筑似给人一种轻盈如水晶石一样的结构。

主体的外形与声学需要和观众席的布置相呼应，玻璃花园大厅及前厅、功能组织良好的侧翼，以及入口雨棚都使该建筑有鲜明特点。

建筑通体由白色瓷砖覆面，建筑的音乐功能通过墙面上的瓷砖和石块相间排列而表现出来。

建筑的框架系统多有创意，如前厅的绿化设计以及主厅顶棚上的植物主题。

连接大厅两个入口的内部通道对入口层及整个建筑而言都形成了功能框架。

办公，会议，银行及邮局都布置在100m长7m宽的内走廊中，由芬兰著名艺术家吉蒙·凯瓦托创作的整个建筑中最重要的艺术作品也布置在此处。

职员用房布置在内走廊靠近观众厅一侧，另一侧则通向开敞的前厅，其玻璃花园大厅则通向索萨普斯特公园，而前厅在层高上高出一些。

前厅的设计是为了创造一个宽敞的公共空间，以便为大量人流提供足够的交流场所，同时也安排有一系列的半私密小角落以及座位区。

两个音乐厅不仅功能不同，而且建筑表现手法也不一样。公开的入口及内走廊允许两个音乐厅分别独立使

'Factory Siren' was the title of the winning proposal chosen from among 72 entries in the open architectural competition in 1983. The title links the plan of the concert and congress hall with industrial Tampere.

The location of the building in a park made it possible to plan it as a completely independent unit. Despite its size, the Hall gives an impression of being a light, crystal-like construction.

The overall appearance of the Hall conforms with the acoustic requirements and the layout of the auditoria. The design is characterized by the glass garden halls and foyers, the functional administrative wing and the canopies of the entrance court.

The body of the building is covered with white ceramic tiles. The musical function of the building is expressed in the musical staves which are in ceramic slab and natural stone on the elevations of the auditoria.

The impressive frame of the building has many surprises, such as the oasis with its greenery and fountains hidden in a jungle of glass metal foyers, and the foliage motif on the ceiling of the main auditorium.

The interior passage linking both entrances of the Hall forms the functional frame of the entrance floor and also for the whole building.

The box offices, congress, bank and postal services are placed along the 100 metre long and 7 metre wide inner passageway. The Hall's most important work of art by Kimmo Kaivanto, one of Finland's best known artists, can also be found here.

The cloakroom facilities are situated on the audience side of the passageway. The other side of the walkway opens onto the airy and spacious foyer area, its glass garden halls opening onto Sorsapuisto Park and the foyer levels above.

The idea of the foyer design was to create spacious public lobbies that provide sufficiently expansive levels to cope with large numbers of people, and at the same time a variety of intimate corners and seating groups.

The two auditoria differ from each other not only in their fucntions, but

1

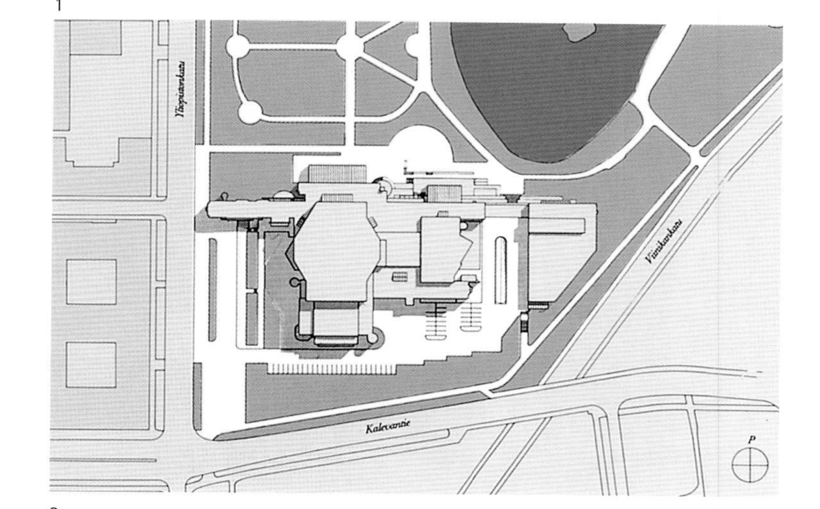

2

3

用。大音乐厅主要用于重要的音乐会，可容纳1870人，其600m^2面积的舞台可用于歌剧、交响乐等大型演出。该音乐厅的2.2秒混响时间对交响乐的演奏是非常理想的。通过某种调节，其混响时间可调至1.5秒，从而也使之适合于开会的功能。

小观众厅作剧院用时可以安排548张座位，其他时候可以布置315张桌子。

它的音响效果按室内乐表演设计，混响时间为1.5秒，但借助一个可调节削弱器，它的混响时间可缩短到演讲所需要的1秒。排演大厅可以用来演奏实验性音乐、录音，也可作为一个音乐实验室，它设有手动的音效控制。

大厅可以容纳165～306位观众，位于两个主要观众厅之间，在大型会议期间，它可以作为便利的会议和讨论厅。它也可以被用作展览室。

除了观众厅之外，该建筑还包含了很多各式各样的会议室、一个可容纳200人同时用餐的标准餐厅、私人进餐室、位于冬日花园的自助餐厅、贵宾休息室和一个总面积为1265m^2的展览空间。

这些设施可以为3000人的不同的大型会议或音乐会提供服务。它既能使各种活动分开独立举行互不干扰，也可以用作单一的目的。

这个建筑物的休息区、舞台区、排演厅和观众厅之间的中心部分的结构是预制梁柱。礼堂的支撑墙结构都是滑模钢筋混凝土。

礼堂的屋顶采用钢结构，空间构架梁和钢筋混凝土构件。休息室的玻璃花园大厅由钢网格结构来支撑的。

also in relation to architectural expression. The separate entrances and the inner walkway allow independent use of both auditoria with their respective foyer areas.

Planned primarily as an important concert hall the main auditorium seats 1870 persons. The 600m^2 stage has been planned for opera, choir and orchestra performances. The auditorium's 2.2 second reverberation time is ideal for orchestral music. By using special constructions, it can be brought down to 1.5 seconds, thereby making it perfect for congress purposes.

The small auditorium seats 548 persons in theatre style and 315 at desks. Designed for chamber music, its 1.5 second reverberation time can be shortened by a modifiable attenuation structure to 1 second for speech.

The rehearsal hall can be used for performaces of experimental music, recordings and as a music laboratory. It has manual acoustics controls.

The hall seats 165-306 persons. Located between the two main auditoria, it is a convenient meeting and seminar room during congresses. It can also be used for exhibitions.

In addition to the auditoria the building includes numerous adaptable conference and seminar rooms, a 200-seat restaurant of gourmet standard, private dining rooms, cafeteria in the winter garden, VIP foyers and a total of 1,265m^2 of exhibition space.

The facilities make it possible to arrange different congress and concert events for up to 3,000 people. The events can be independent of each other or the whole building can be used for a single function.

The main section of the foyer areas, the stage area, the rehearsal hall wing and the central part of the building between the auditoria are based on a pre-fabricated column-beam system. The supporting wall structures of both auditoria are of slip-cast reinforced concrete.

The roof structures of the auditoria are of steel, space frame beams and reinforced concrete elements. The glass garden hall of the foyers are supported by a steel lattice construction.

4

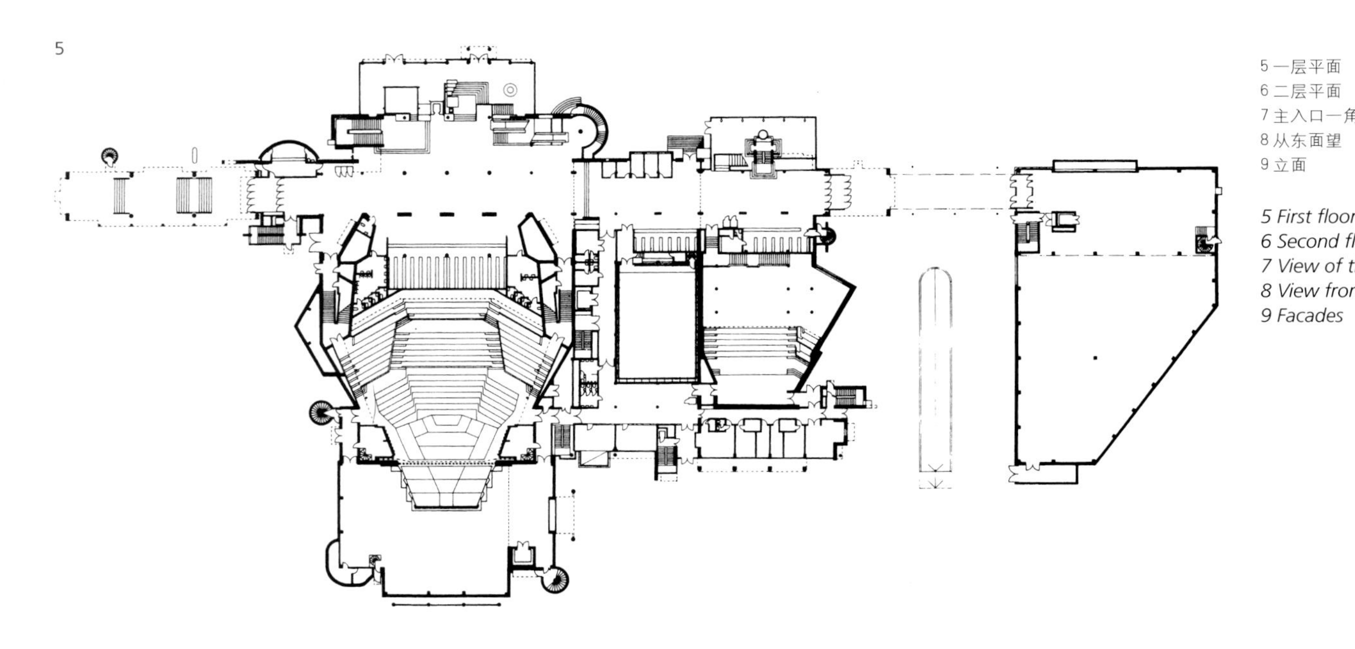

6

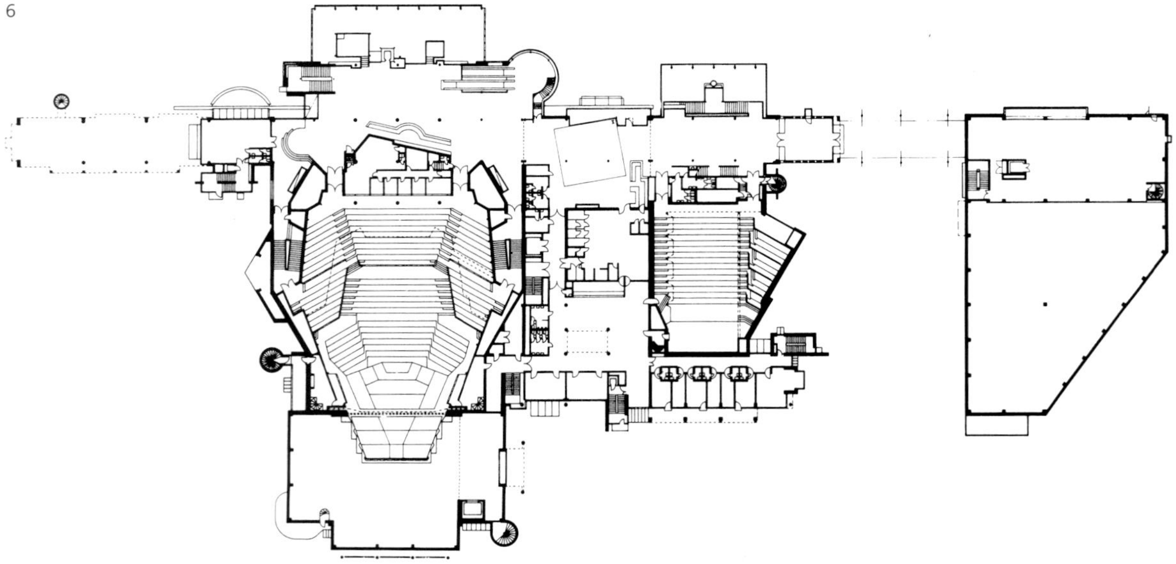

5 一层平面
6 二层平面
7 主入口一角
8 从东面望
9 立面

5 First floor plan
6 Second floor plan
7 View of the main entrance
8 View from the east
9 Facades

7

8

9

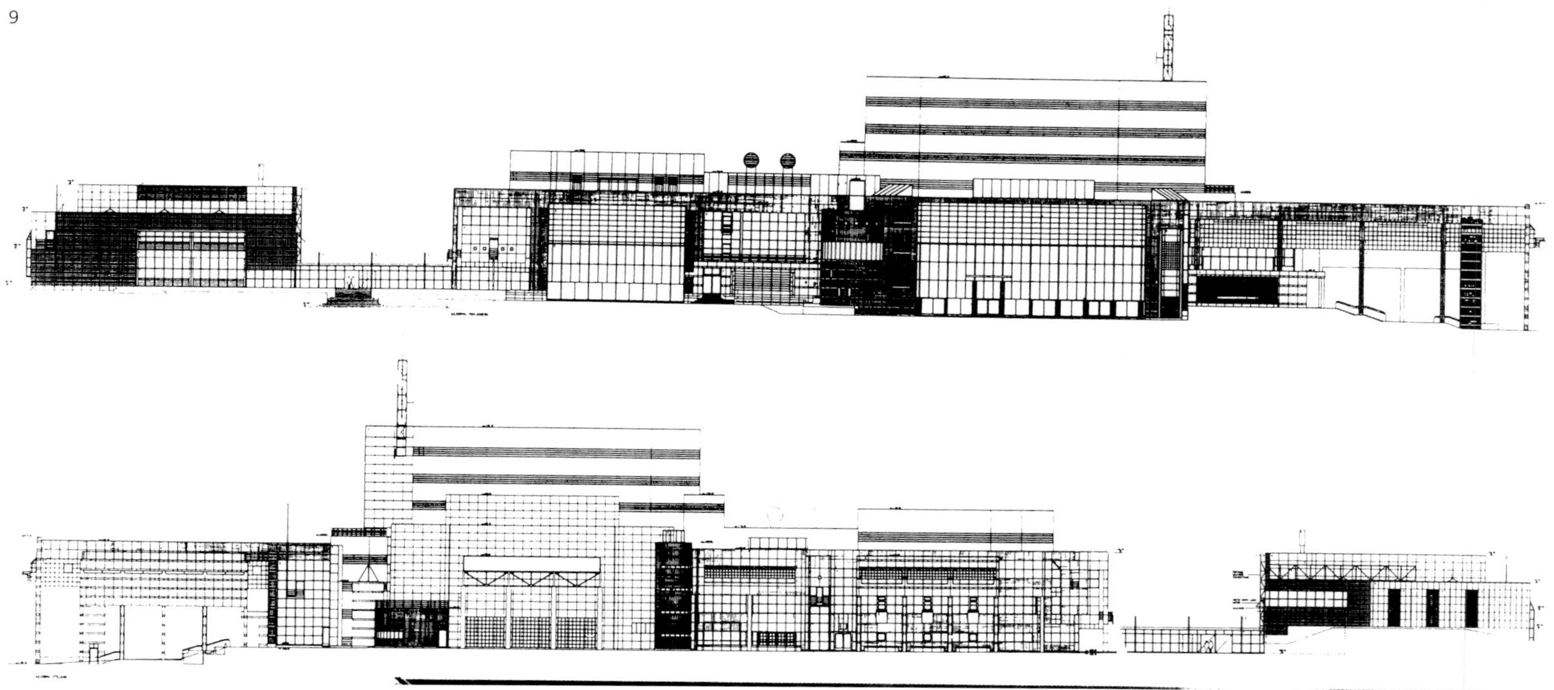

10 剖面
11 主休息厅透视
12 喷泉池平面
13 室内瀑布及喷泉轴测图
14 主休息厅咖啡室
15 大观众厅

10 Section
11 Main foyer perspective
12 Fountain basins plan
13 Water cascade and fountain axonometric
14 Main foyer cafe
15 Large auditorium

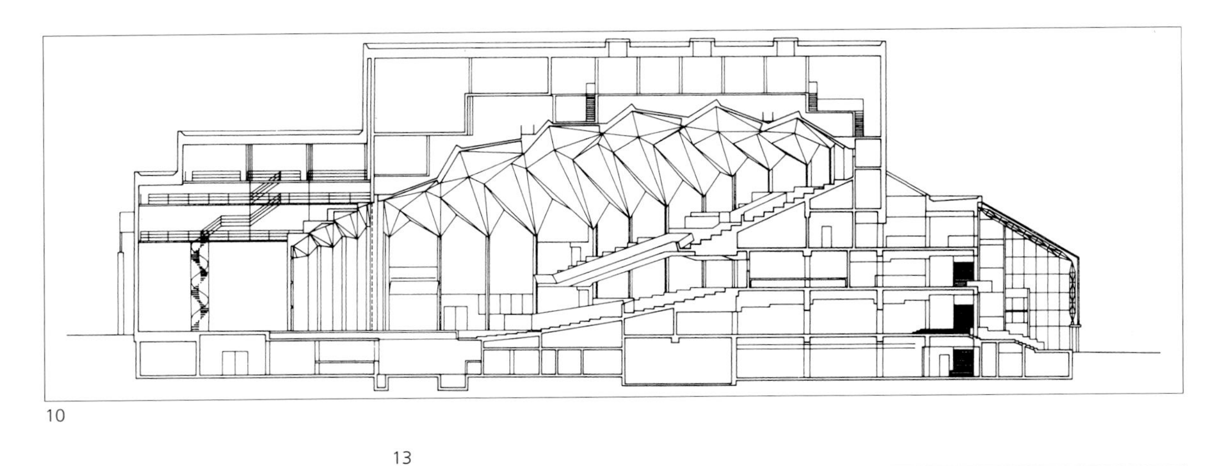
10

11

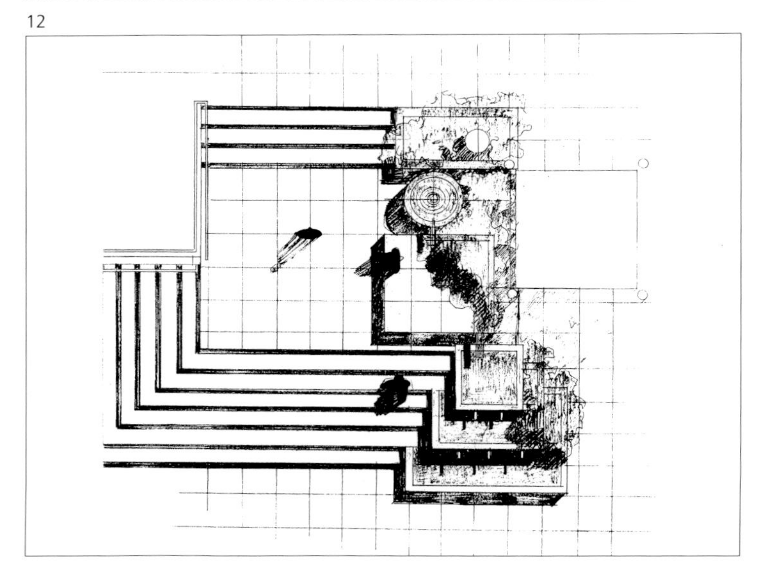
12

13

16 大观众厅剖面
17 小观众厅
18 大观众厅
19 大观众厅剖面
20 小休息厅
21 展览厅入口
22 从大观众厅舞台上望
23 从公园看的夜景
24 雨篷
25 室内细部

16 Large auditorium section
17 Small auditorium
18 Large auditorium
19 Large auditorium section
20 Small auditorium foyer
21 Exhibition hall entrance
22 View from the large auditorium stage
23 View from the park at night
24 Canopy
25 Interior detail

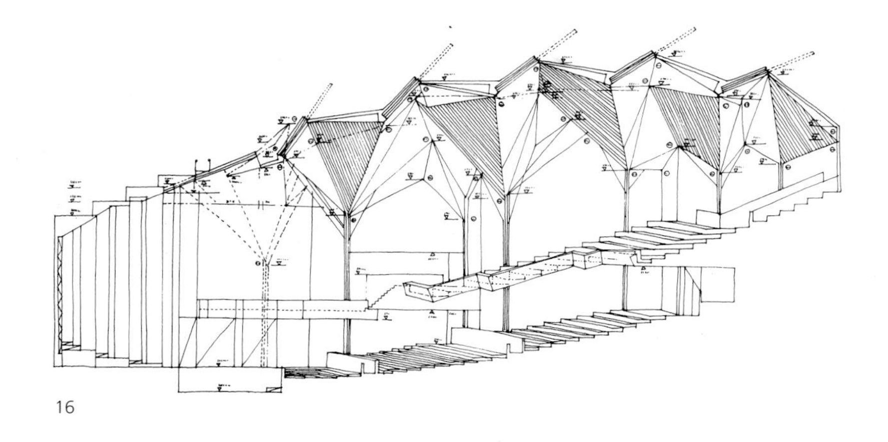

16

17

18

19

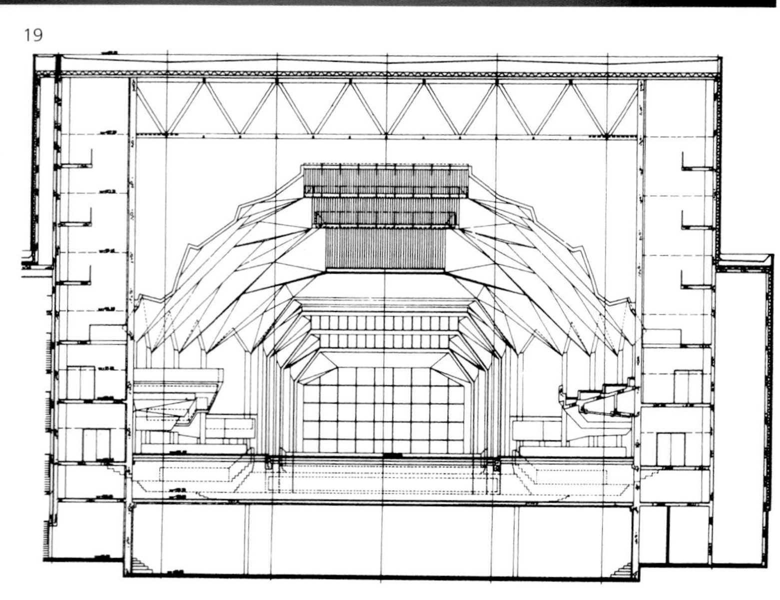

21

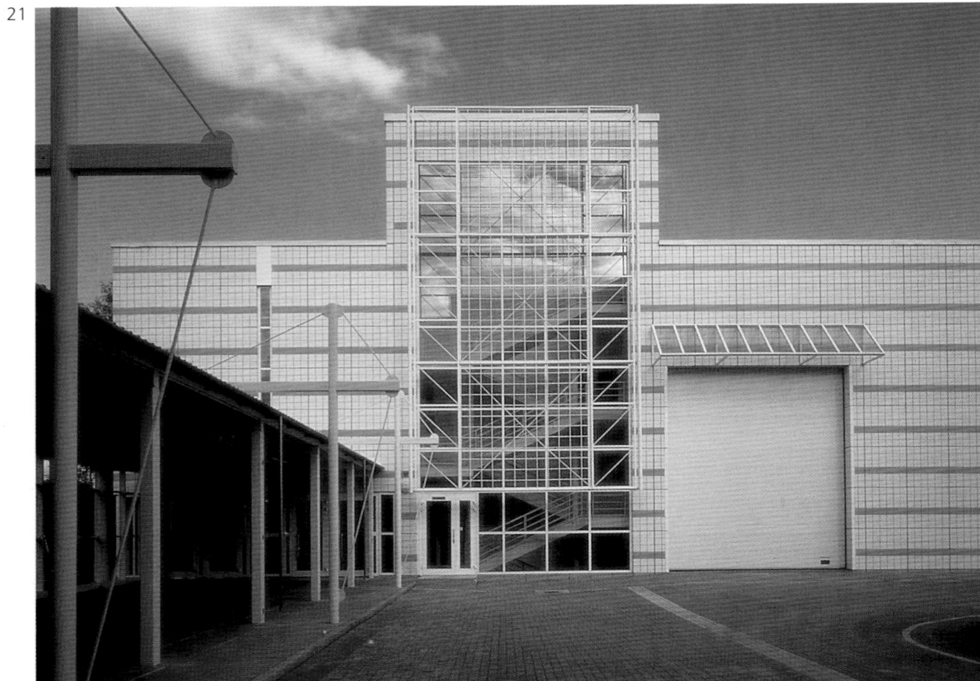

22

23

24

25

劳季奥办公大楼，埃斯波 1990

Office Building Rautio, Espoo 1990

1 总平面
2 二层平面
3 一层平面
4 地下层平面
5 模型照片

1 Site plan
2 Second floor plan
3 First floor plan
4 Basement plan
5 Model photo

大楼包括一系列的办公室，其中有些是开敞式平面，另一些是小的隔间，除此之外是必要的员工和社会的设施。设计任务书也要求考虑公司内部培训设施的要求，员工交往空间的要求，以及桑拿室附近客房服务的要求，并希望室外空间在举行员工活动时可以被利用起来。营造的舒适工作环境是设计的重点所在。此外，需要一个高质量的建筑来突出公司的形象。

根据城市规划要求，建筑为两层。地形是一个自街道向建筑的后部降低的斜坡，使得一层位置可以通过玻璃面向外开敞，获得更多的自然光。

建筑的中心部分是一个两层高的预制结构的办公区。在这一结构内有一个联系上下两层的采光井。在基本结构之外是建筑富有特色的钢结构部分：比如在街道面的入口处理，东南方向的遮阳设施，消防梯，采光井的玻璃结构等等。细部设计希望获得一种具有图形感的、技术精美的效果。建筑的金属外表面有不锈钢，经热镀处理的面板，涂漆面板。金属的不同质感和面层处理反映了建筑内部不同的使用功能。

The building was to include a number of offices – some open-plan, others cubicle-like – and the necessary staff and social facilities. The brief also specified facilities for internal company training sessions, gatherings of all the staff, some of whom working outside the company and guest facilities abutting the sauna suite. It was hoped that space outside would be accessible during staff events. Staff satisfaction and good working conditions were particularly emphasized. In addition, the aim was a building of high architectural standard which would reinforce the company's overall image.

The building has two floors as specified in the town plan. The sloping lot, which descends from the street towards the rear of the building, allowed the ground floor to open outwards with a glass enclosure, so that the staff facilities could have natural light.

The core of the design is a two-storey prefabricated office framework. This frame contains a skylight linking the two levels. Leading off the basic framework are the steel-built identity components of the building: the lobby at the street end, the sun-shield structures to the southeast, the emergency stairs, and the glass-roofed structures of the skylight with their advertising surfaces. The detailing was intended to achieve a graphic, technically polished inpression. Some of the building's steel surfaces are stainless, others have been hot-galvanized and others are painted. The various qualities of steel and surface treatments are intended to emphasize the different purposes of the various parts of the building.

1

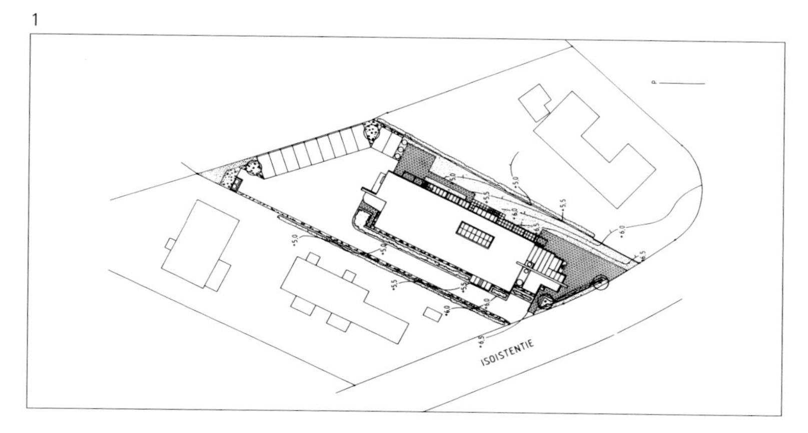

2

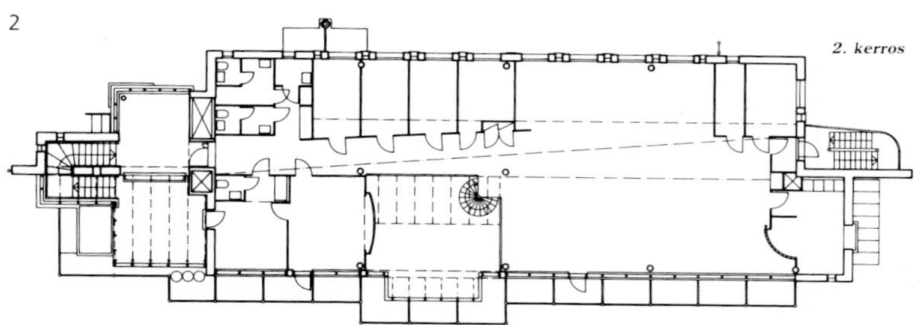

3

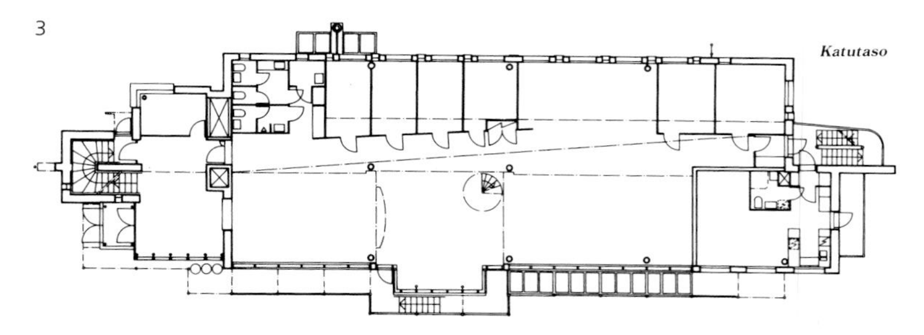

4

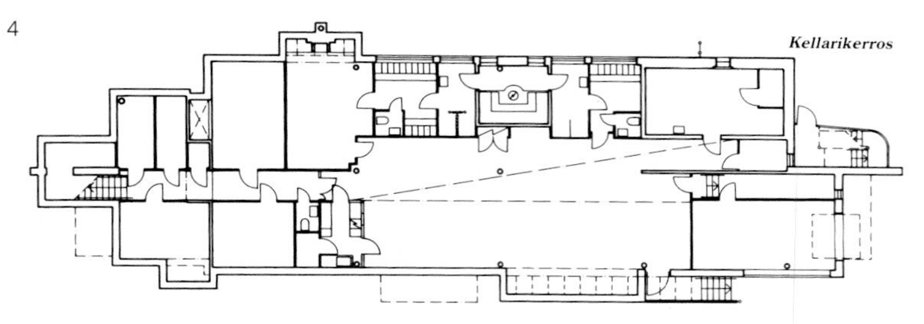

5

6

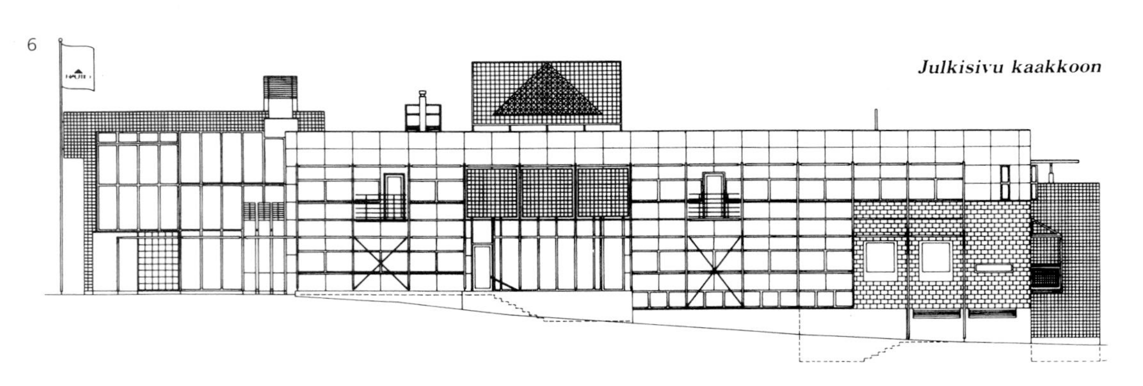

7

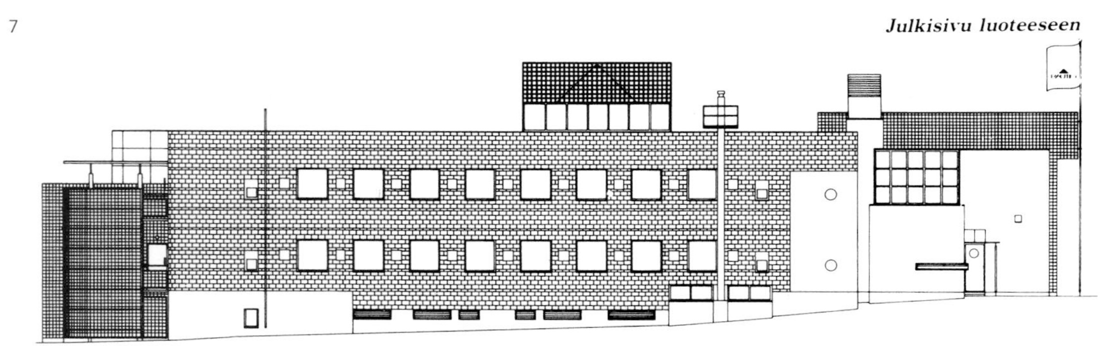

8

9

6 东南立面
7 西北立面
8 细部
9 温室
10 主入口

6 South-east facade
7 North-west facade
8 Detail
9 Greenhouse
10 Main entrance

RAUTIO

11

12

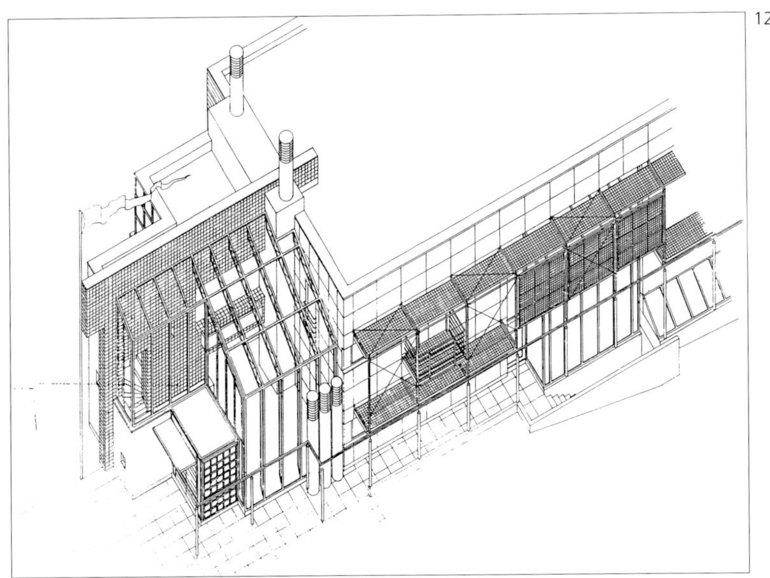

13

14

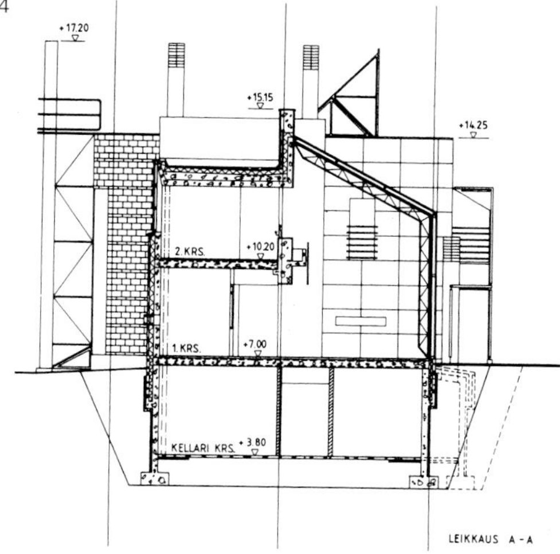

15

16

11　透视
12　轴测图
13　外观
14　剖面
15　室内
16　楼梯

11　Perspective
12　Axonometric view
13　Exterior view
14　Section
15　Interior view
16　Staicase

维基·特里安琪尔公墓，设计竞赛第一名，赫尔辛基 1990

Viikki Triangle Cemetery Competition entry, 1st prize Helsinki 1990

公墓的主要元素是一个狭窄的楔形中心通道。

从大门入口开始，是一条石板铺成的“石径”，引导人们通向最终的主题。这些东西象征着基督教朴素的朝圣。

从中心通道的两边是修剪得很高的白杨树篱。开敞明亮的墓区后面是茂密的绿阴下的“小教堂”，这里用来安放骨灰瓮。建筑和墙体构成了这个地区的主要入口。

建筑和墙限定出公墓的主要入口。和轻质结构相比，墙象征着永恒。

The main element of the cemetery is a narrow wedge-like centre aisle. Starting from the main gate, this is a 'stone way' featuring slabs with carved sentiments, leading to the final motif. Together they symbolize the Christian's earthly pilgrimage.

From the central aisle narrow side aisles edged with high-trimmed poplar hedges open out. These spacious, light-filled burial units lead to 'chapels' sheltered by a close-growing group of trees and used for urn burials.

The buildings and their walls form the area's main gateway. Being a permanent construction on which the lightly constructed buildings used for more mundane purposes are supported, the walls symbolize permanence.

1

2

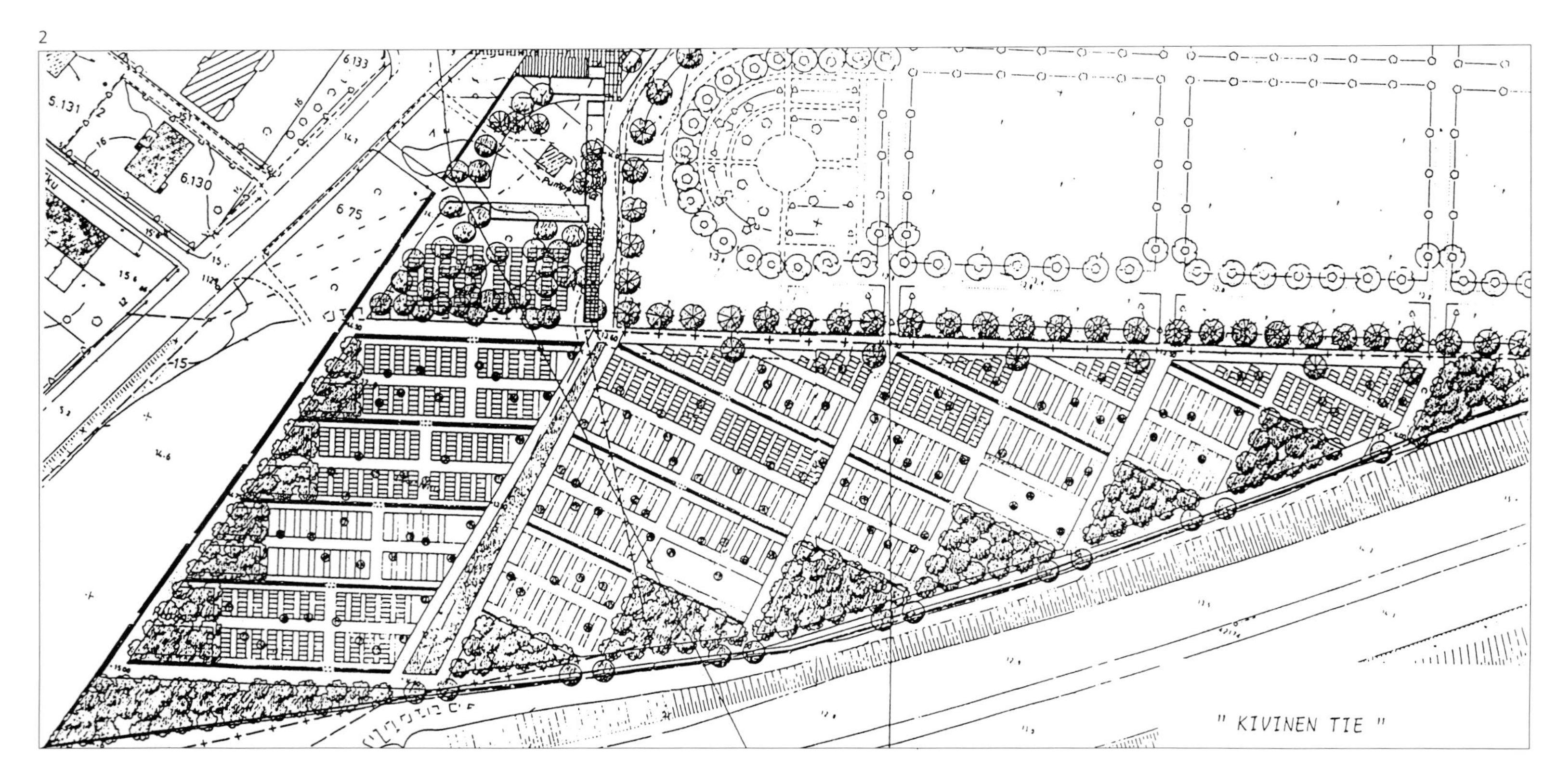

1 主入口透视
2 总平面
3 中央通道透视

1 Main entrance perspective
2 Site plan
3 Centre aisle perspective

3

4 中央通道
5 宁静的墓园

4 Centre aisle
5 Silence of the cemetery

5

4

考哈约基国内经济学校
1992

Kauhajoki School of Domestic Economics Kauhajoki 1992

扩建部分的建筑紧挨着老校舍，老校区的建筑仍然在环境中占主导地位。新的大楼按功能分成几个部分。一个高耸的礼堂占据了新楼的主要部分。礼堂的空间可以通过一面滑动的墙与餐厅结合或隔离开来。礼堂的四壁是玻璃墙，中央是明亮的空间。

为不同教学目的服务的设施被放在新大楼的侧翼。礼堂的设计也考虑了音质效果。

新大楼里有教学用的厨房和餐厅、用于教授清洁技能教室、织物保养室、教学用洗衣房、社会服务室、带餐厅的大型厨房、礼堂和图书馆。

新大楼一部分为现浇，另一部分为构件组装。大楼的支撑柱是钢筋混凝土结构，结合了一部分木结构。

木屋架用的是预制构件。建筑物的立面是白色的砖和漆成白色的板。内部的颜色也较淡。

老大楼也在充分尊重其历史价值的基础进行了翻新。

The annex is placed next to the old school building in a way that the old building still dominates the scenery. The new building has been divided into functional sections. A high auditorium dominates the new building. The auditorium can be linked to the adjoining dining room and hall with a sliding wall. The hall has glass walls and a light central space.

The facilities serving the different teaching purposes are in the wings of the annex. Also musical requirements have been taken into account in designing the acoustics of the auditorium.

In the annex there are teaching kitchens of domestic economics with dining areas, rooms for cleaning techniques teaching, textile care rooms, teaching laundry, rooms for social services, large-scale kitchen with dining rooms, auditorium and library.

Parts of the annex have been built on the site, parts from elements. The supporting pillars are concrete and parts on the first floor wood.

The wooden roof grilles are built of elements. The supporting pillars are concrete and on the first floor wood as well. The facades are white brick and board painted white. The interior colours are light.

The old main building has also been renovated with respect of its historical values.

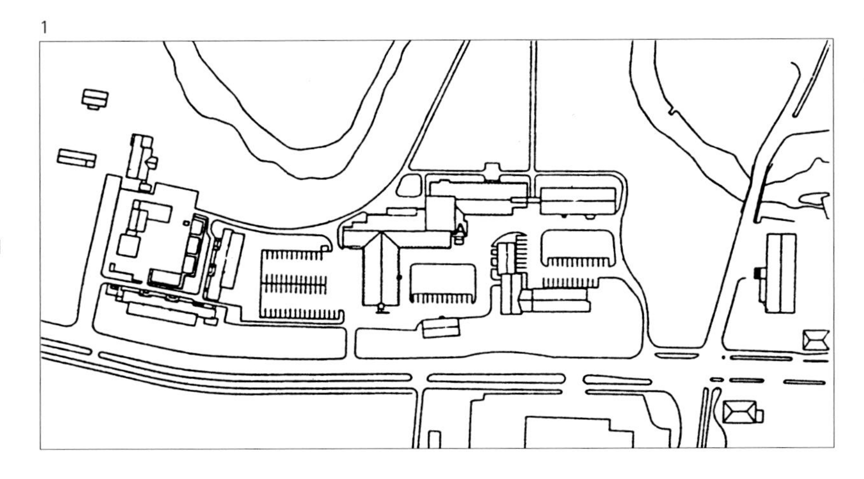

1 总平面
2 底层平面
3 二层平面
4 主入口

1 Site plan
2 Ground floor plan
3 First floor plan
4 Main entrance

5 辅助入口
6 从河上望
7 从河上望
8 夜晚的主入口
9 与学校旧建筑的连接
10 主入口大厅

5 Secondary entrance
6 View from the river
7 View from the river
8 Main entrance at night
9 Connection to the old school building
10 Main entrance hall

5

6

7
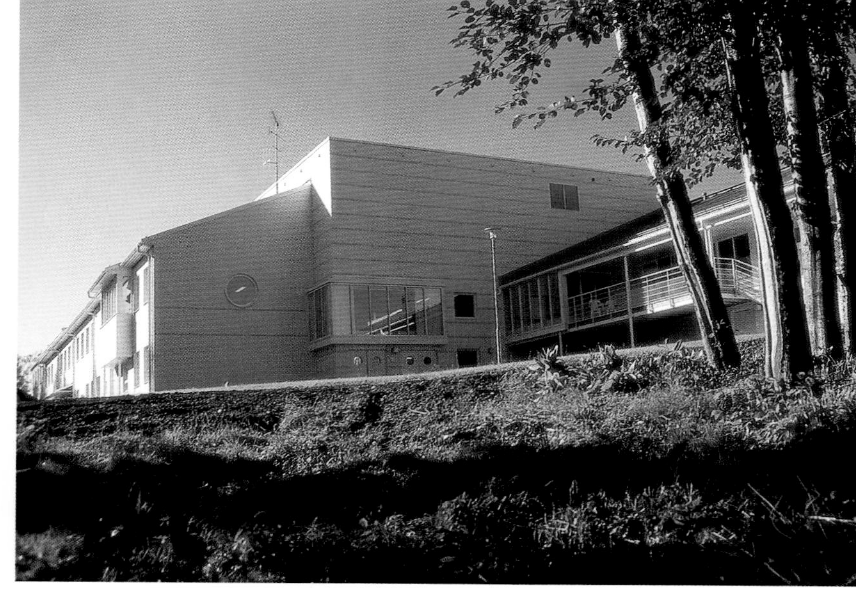

8

9

11　主入口大厅
12　观众厅

11　Main entrance hall
12　Auditorium

11

12

塞内约基火车站，隧道和站台 1993

Seinäjoki Railway Station, Passenger Tunnel and Platform Shelters 1993

芬兰国家铁路局正在为适应未来的高速列车的需要进行全国性的车站更新。

塞内约基火车站建成于1971，是当时理性建筑的代表。在车站里必须穿越轨道才能到达不同的站台。高速列车的引进使车站有必要比原计划提前建造一条人行隧道。同时，期待已久的站台的顶棚也建造了起来，尽管由于预算紧张使建造的顶棚面积被控制到最小。隧道像平板桥一样建造在钢管桩基上，周围是隔热挡土墙。内表面覆盖了线路板，来自背后不同色调的灯光制造出一种“光的艺术”。

楼梯顶和站台遮阳棚采用了钢和玻璃的结构。电梯井的结构也是钢和玻璃。钢构件是在工厂预制并在现场组装的。玻璃采用6+6mm的钢化压合玻璃。钢是表现交通建筑特点的理想材料。钢和玻璃相结合制造出通透这一交通建筑特有的效果。玻璃技术的革新是使得玻璃建筑的复兴变成可能。

这类旨在为芬兰国家铁路局树立新形象的建筑项目同时也创造出新的商业机会。

1 总平面
2 剖面

Finnish State Railways is currently renovating its stations all over the country for future high-speed trains.

Seinäjoki Station was completed in 1971 and represents the rational architecture of the time. The different platforms have been accessible only by crossing the tracks. The introduction of the new high-speed train made it necessary to start the construction of an access tunnel sooner than originally planned. At the same time, platform shelters, long waited for, were built, although the tight budget kept the number of shelters at a minimum.

The tunnel has been built on steel pipe piles as a slab bridge with heat-insulated and shot-crete-steel piling walls as earth pressure walls.

The inside walls are covered with wire plate cassettes and the space behind them is illuminated in different shades to create 'artwork of light'.

The stair roofings and the platform shelters are of steel and glass construction. The structures of the elevator shaft are also of steel and glass.

The steel sections are prefabricated at the factory and assembled on the site. The glass used is 6+6mm tempered and laminated glass. Steel is ideal for the character of traffic centres. Together with glass it creates a transparent effect that can be considered one of the basic elements of a traffic centre's architecture. The tehcnical innovations connected with glass has made a new renaissance of steel-glass architecture possible.

This type of architecture supports Finnish State Railways in its attempt to refresh its visual image now that it has become a business enterprise.

1 Site plan
2 Section

1

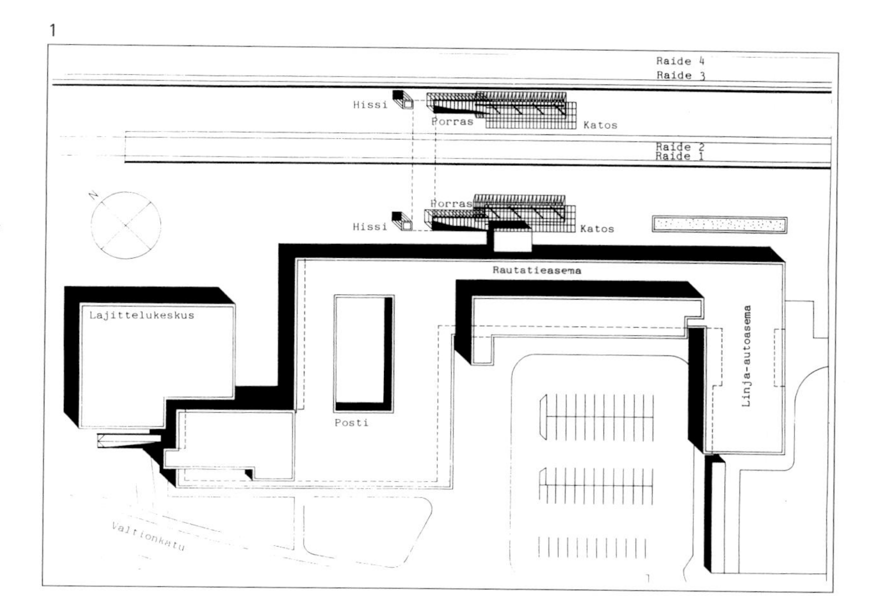

2

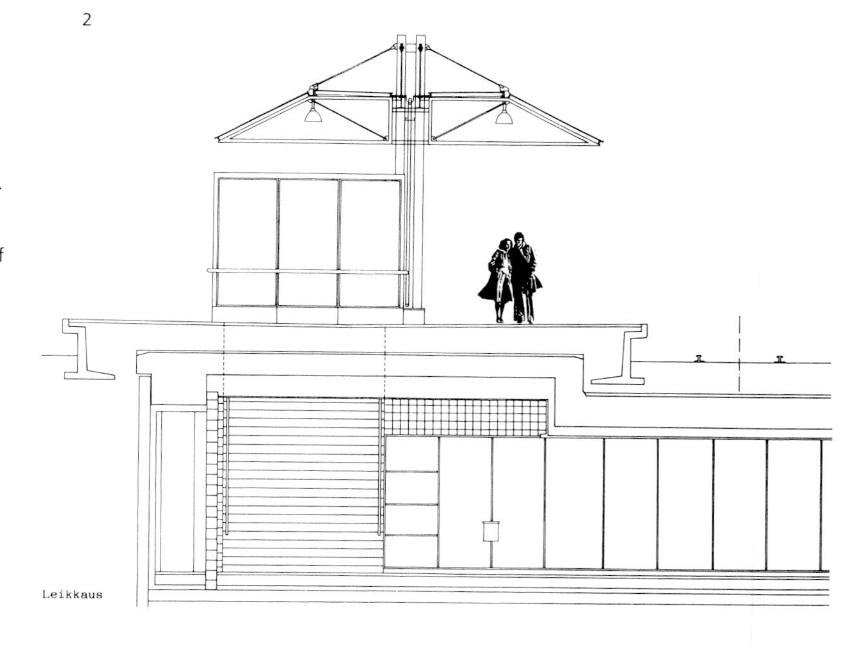

3 东北立面
4 楼梯上方的玻璃盖
5 细部
6 细部
7 月台遮棚

3 North-east facade
4 Glass covered staircase
5 Detail
6 Detail
7 Platform shelters

3

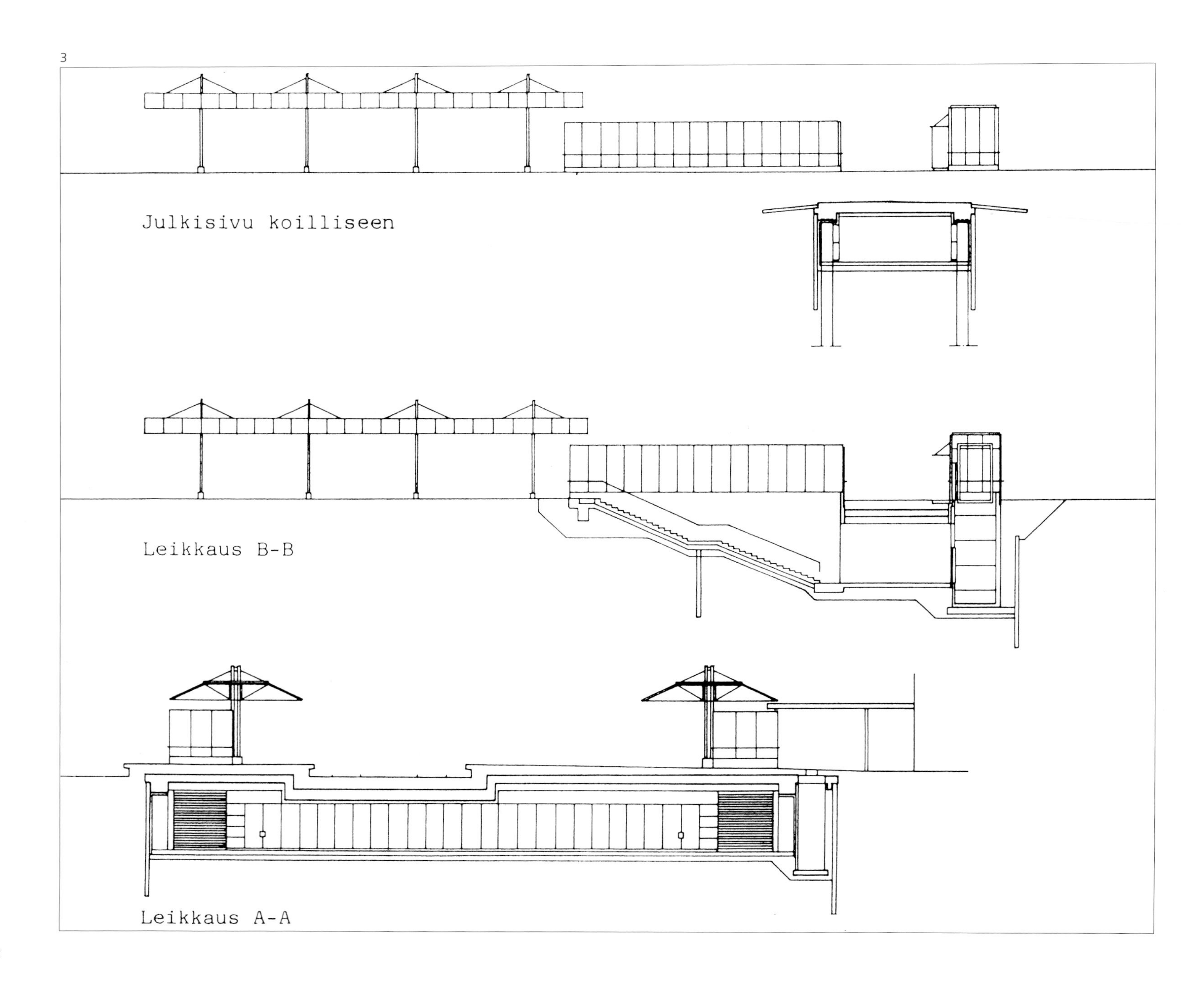

5

6

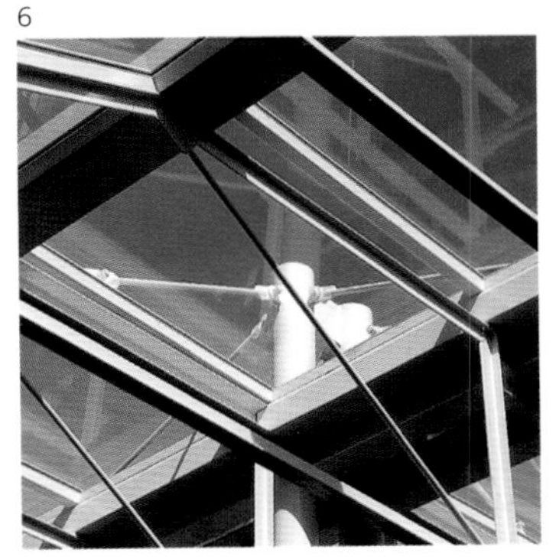

7

8 楼梯
9 月台遮棚

8 Staircase
9 Platform shelter

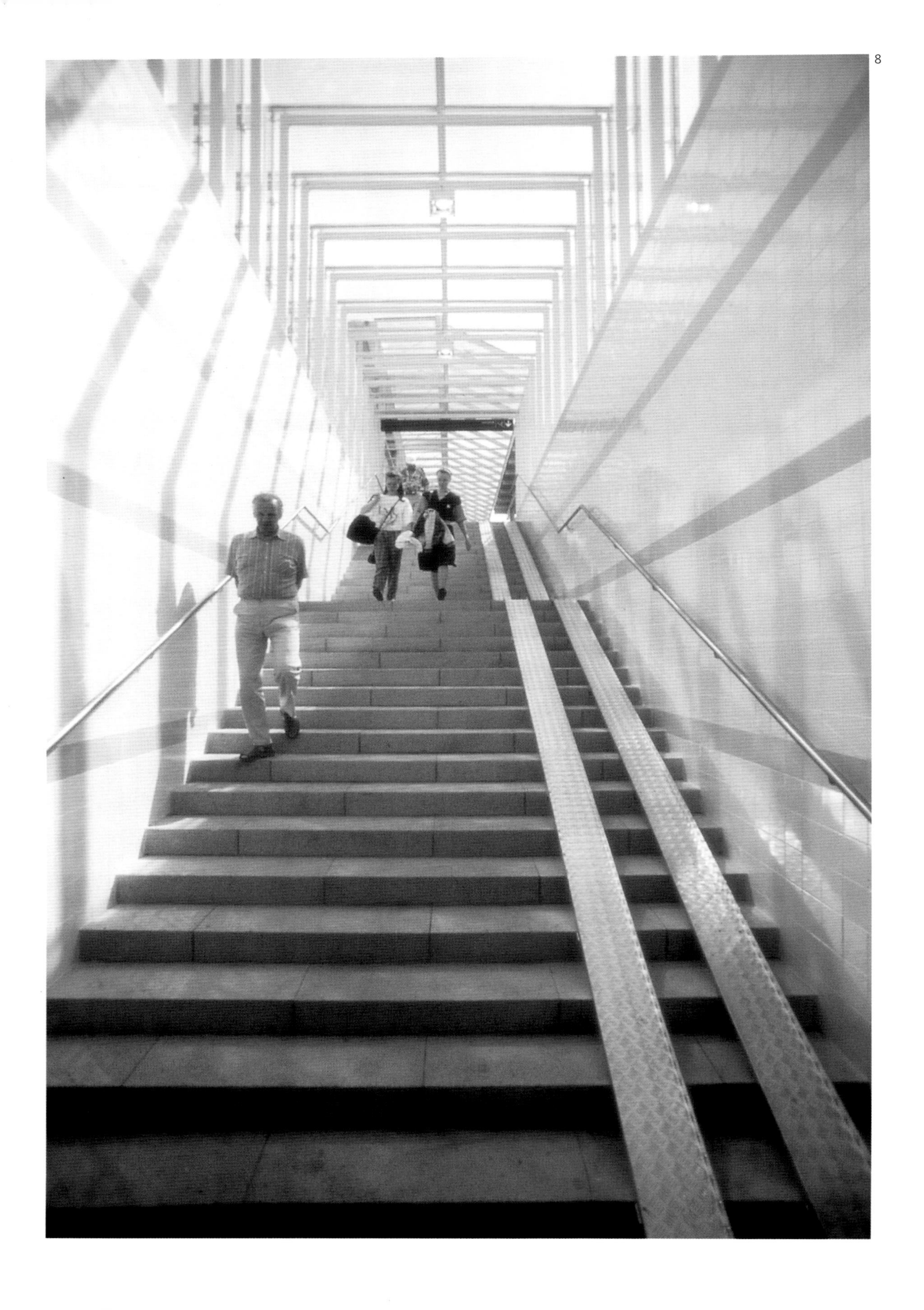

8 9

3

卢纳米住宅区，
莱瑟米
设计竞赛第一名
1993

Luuniemi Housing Area
Iisalmi
Competition entry, 1st prize
1993

在规划这个住宅区的时候已经对附近街区的木屋住宅进行了调查。设计所采用的尺度和开放式的模式就来自当地的传统，而建造技术和建筑式样却是现代的。

在这个地区建造的住宅为2～3层的公寓；沿街的首层是工作室、商业空间和公众集会场所。开阔的庭院可以吸引居民一起来玩游戏、开夏日晚会和展开园艺活动。

In the dimensioning of the housing blocks the local scale of wooden house blocks has been observed. The scale and open building mode are reminders of this tradition, but the construction tehniques and architecture are modern.

There are two and three-storey apartment houses along the area; on the ground floors along the streets there are workshops, business spaces and rooms for public gatherings. The open courtyard invites the inhabitants for common purposes like games, summer parties and gardening.

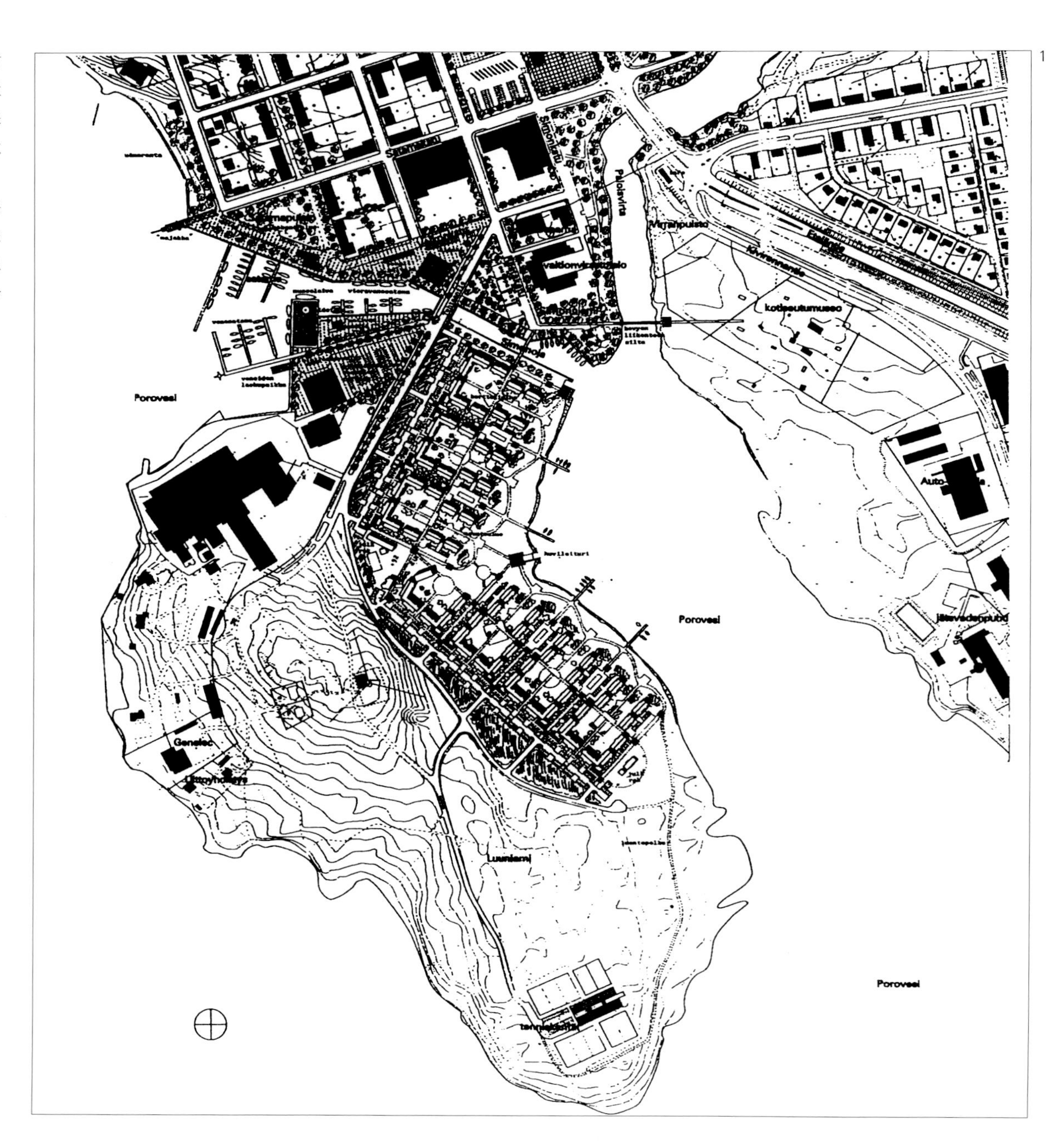

1

1 总平面
2 通往住宅单元内庭院的大门
3 局部平面
4 立面

1 Site plan
2 Gateway to the housing unit courtyard
3 Detail plan
4 Facades

5 庭院内景
6 海边（湖边）

5 Courtyard view
6 Waterfront

凯萨涅米地铁站，站台大厅，赫尔辛基 1995

Kaisaniemi Metro Station, Platform Area, Helsinki 1995

1 赫尔辛基市中心的地铁线
2 候车平台平面

1 Metro line in the centre of Helsinki
2 Platform area floor plan

Kaisaniemi 地铁站代表了新一代的地铁建筑设计。旧的内部装饰继续保留，有一部分作了新的改动。

kaisaniemi 地铁站的设计构思之一就是要把地面上的氛围带到地下来。各种各样的灯光反射在混凝土喷涂墙面上，创造出比实际更宽敞的空间效果。

站台边上的墙面仿照商店的橱窗，采用了凸窗和发光广告。墙体的结构材料是经喷漆处理过的钢架，外面覆盖以玻璃、网格板或穿孔钢板。

站台的顶棚保护了所有的必要设备。顶棚是钢框架，外覆网格板，不但有利通风，而且在上空产生一种浮云的感觉。自动扶梯通道的尺寸按照已挖的隧道设计。与以前的计划不同的是增加了一个对角方位的自动扶梯。略显封闭的自动扶梯通道使用了反光材料，墙面使用的是镜面玻璃，天篷用的是氧化铝面板。

Kaisaniemi Metro Station represents a new generation of metro architecture design. Old interior decorating components are still used but partly with new expressions.

In Kaisaniemi the aim has been to bring some of the atmosphere on street level under ground. Various light effects are reflected on the shotcreted wall surfaces to create the impression of a larger entity.

The walls on the platform's side resemble shop window facades with bay windows and illuminated advertisements. The material of the walls are mainly painted steel framework covered with glass panes, mesh plates or perforated steel plates.

The ceiling of the platform area protects all the required installations. The ceiling is built of mesh plates on a steel frame, making it airy and cloud-like.

The dimensioning of the escalator shaft was based on the already excavated shaft opening. Unlike the previous plans, one diagonal elevator has been added to the shaft. The outside of the rather intimate escalator shaft has been opened up with reflective surfaces; mirror glass on the walls and bright anodized aluminium on the ceiling.

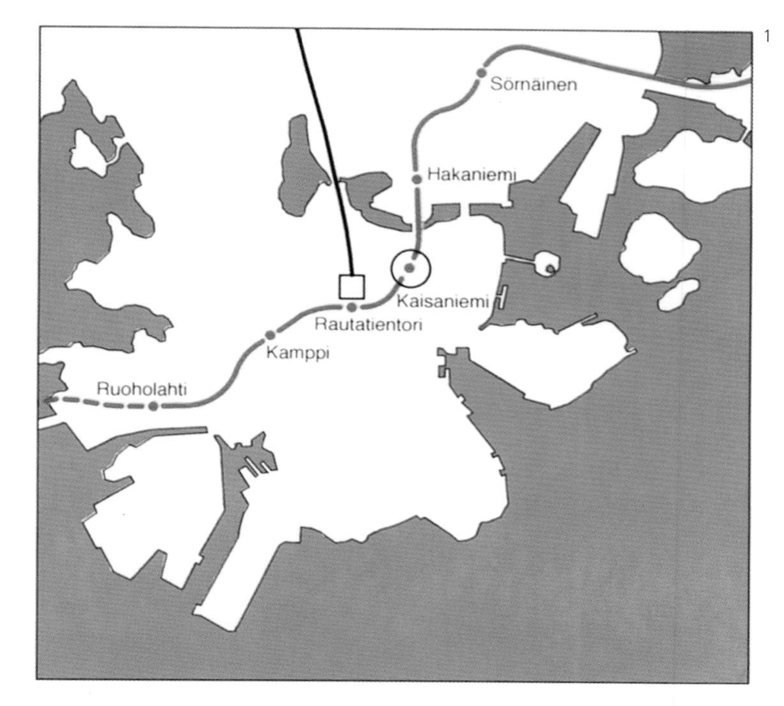

1

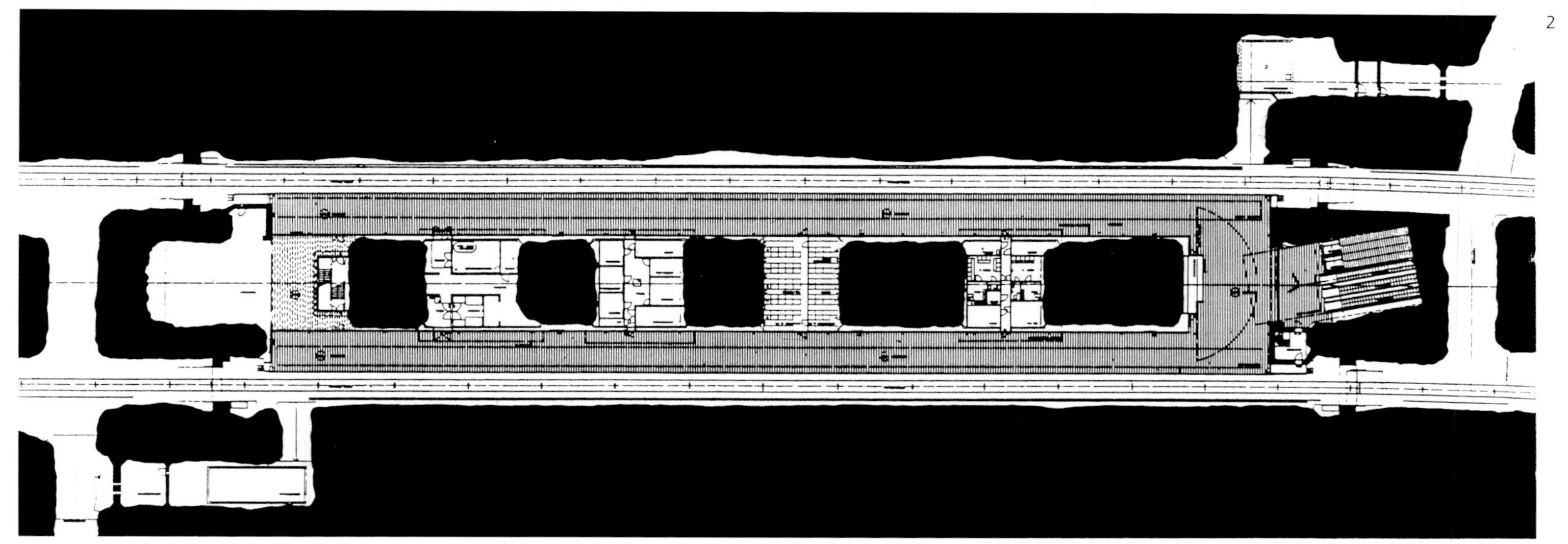

2

3 总平面
4 自动扶梯通道中Annikki Luukela创作的艺术品

3 Site plan
4 Escalator shaft art work by Annikki Luukela

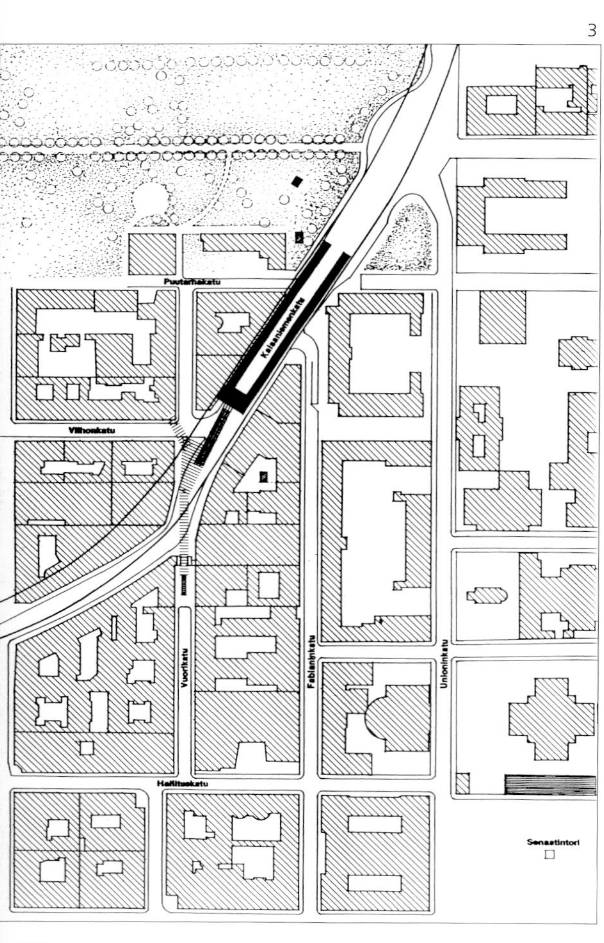

3

4

5

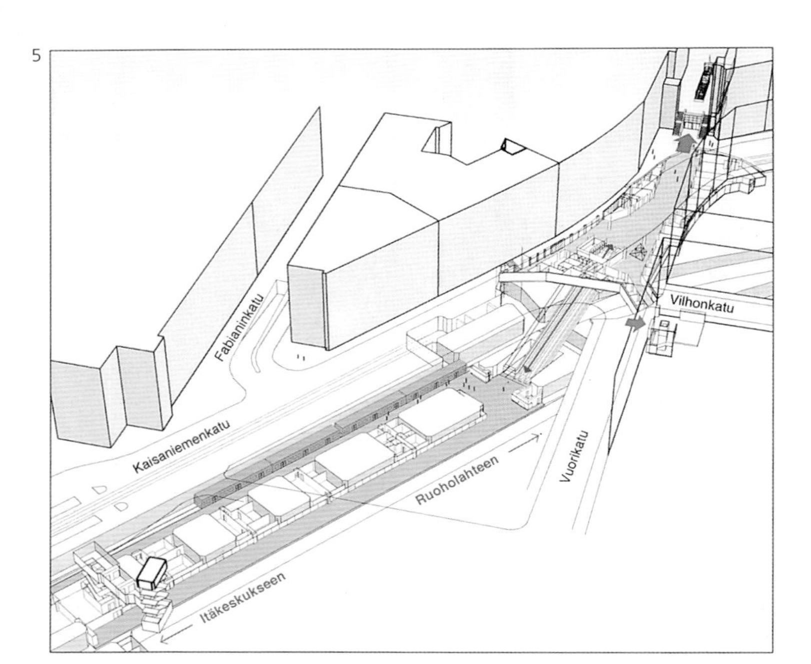
Fabianinkatu
Kaisaniemenkatu
Vilhonkatu
Vuorikatu
Ruoholahteen
Itäkeskukseen

6

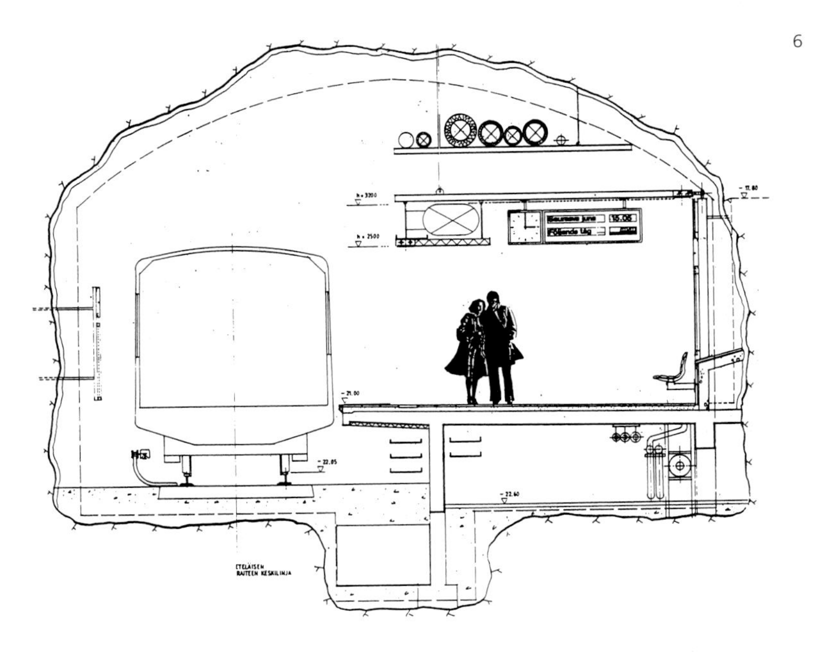

7

Kaisaniemi
Ulos
Ut

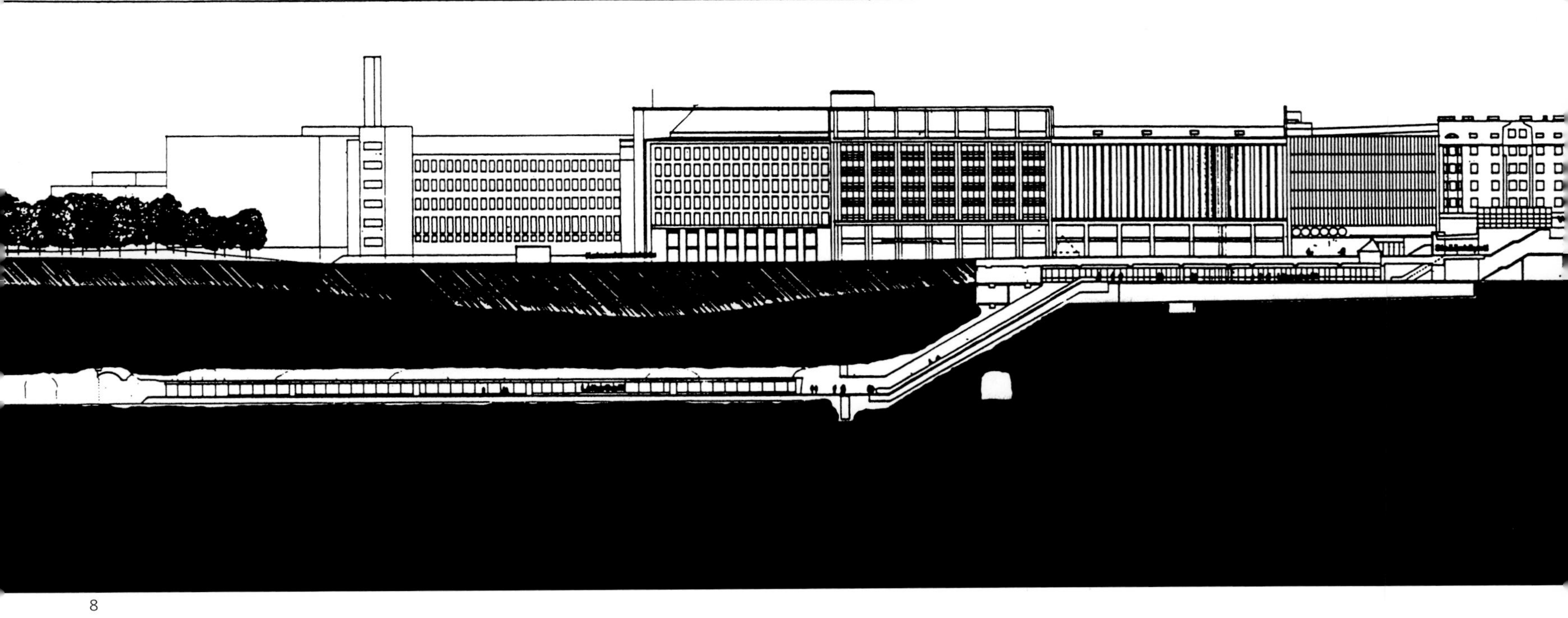

8

9

5 轴测图
6 剖面
7 候车区域
8 剖面
9 候车大厅
10 自动扶梯

5 Axonometric view
6 Section
7 Platform area
8 Section
9 Platform lobby
10 Escalator shaft

11

12

11 自动扶梯
12 自动扶梯门厅

11 Escalatos shaft
12 Escalator lobby

皮库－霍帕莱提
多功能大厅
赫尔辛基 1997

位于端头的单层建筑部分容纳了一个幼儿园和青年活动室，中间部分是一个两层楼的小学。学校的永久性教室在二楼，由两个单元组合在一起。

大楼的分块布局反映出其内部不同的活动功能。各部分之间的分割节点在建筑体量上稍作凹陷。建筑的临街面是小窗户和较厚重的墙，看起来显得很坚固，而朝向庭院一方则较为开放，使得自然光线可以落到建筑的内部。

正立面上有大量的钢制小构件，比如出挑的屋檐、天篷、网眼和穿孔的金属盒和展览架，所有的这些都衬托了整个大楼的外表。空调设备放置在幼儿园的屋顶，即体育馆的后面。体育馆的立面由薄片金属夹板单元制作。另外，空调设备上面覆盖了用机械方式封合起来的金属板。青年活动室的内墙使用声学板，外罩多孔金属板。体育馆和青年活动室的顶棚用的是表面经过喷粉处理过的焊接防护网。

在室内，建筑的金属感通过中央大厅中的钢铁构件，电梯井、开放式钢楼梯以及在二层部分连接学校一端的桥很好地反映出来。

1 早期草图
2 位置图
3 总平面

1 Early sketch
2 Situation plan
3 Site plan

1

2

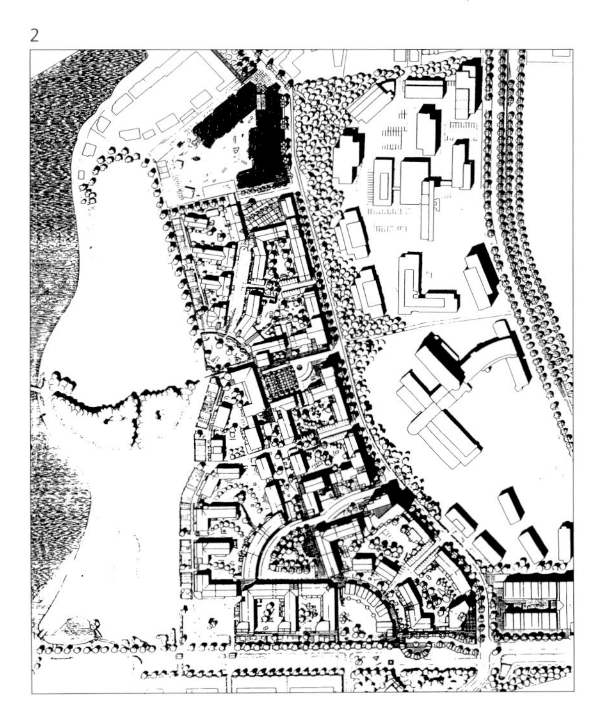

3

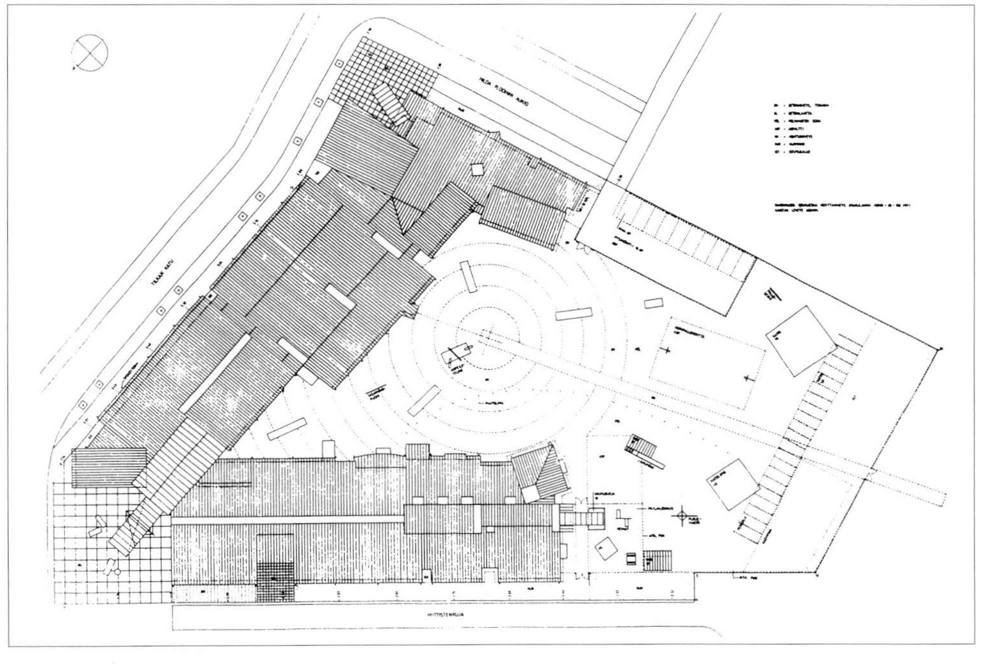

Pikku-Huopalahti Multipurpose Hall Helsinki 1997

4 二层平面
5 一层平面
6 主入口天篷

4 Second floor plan
5 First floor plan
6 Main entrance canopy

The one-storey ends of the building house a kindergarten and youth facilities with a two-storey primary school between them. The permanent classrooms of the school are on the second floor, grouped in two cells.

The various sections of the building reflect the activities going on inside them. The partition points between the sections are somewhat recessed from the actual building masses. The street-side façade of the building is more solid, with small windows and heavy walls, while the fa?ade opening to the courtyard is more open, allowing natural light to the interior of the building.

The numerous steel details on the façade surfaces, such as projective eaves, canopies, meshed and perforated plate cassettes and display cases, define the general appearance of the building. The air conditioning plants are located on the kindergarten roof and behind the gym. Tilkankatu façade of the gym is made of thin sheet metal sandwich units. In addition, the air conditioning plants are clad with mechanically seamed sheet metal. The cladding on the facades of the youth premises consists of hot-dip galvanized steel sheeting, fastened with capnuts. The inside walls of the youth premises are built of acoustic boards covered with perforated metal panels. In the gym and youth premises the ceiling is made of welded protective mesh that has been powder-coated.

The steely nature of the interior is best reflected in the steel-constructed elements of the central lobby, the lift tower, the open steel staircase and the bridge that connects the wings of the school on the second floor.

4

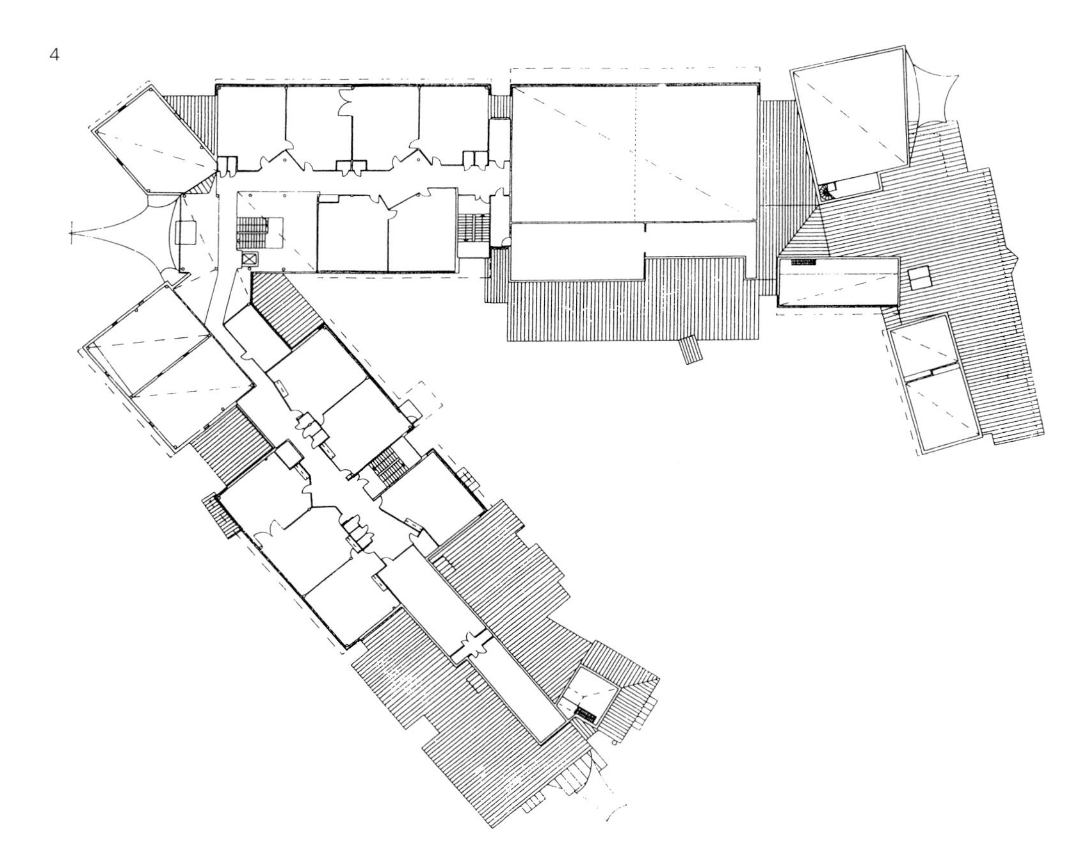

5

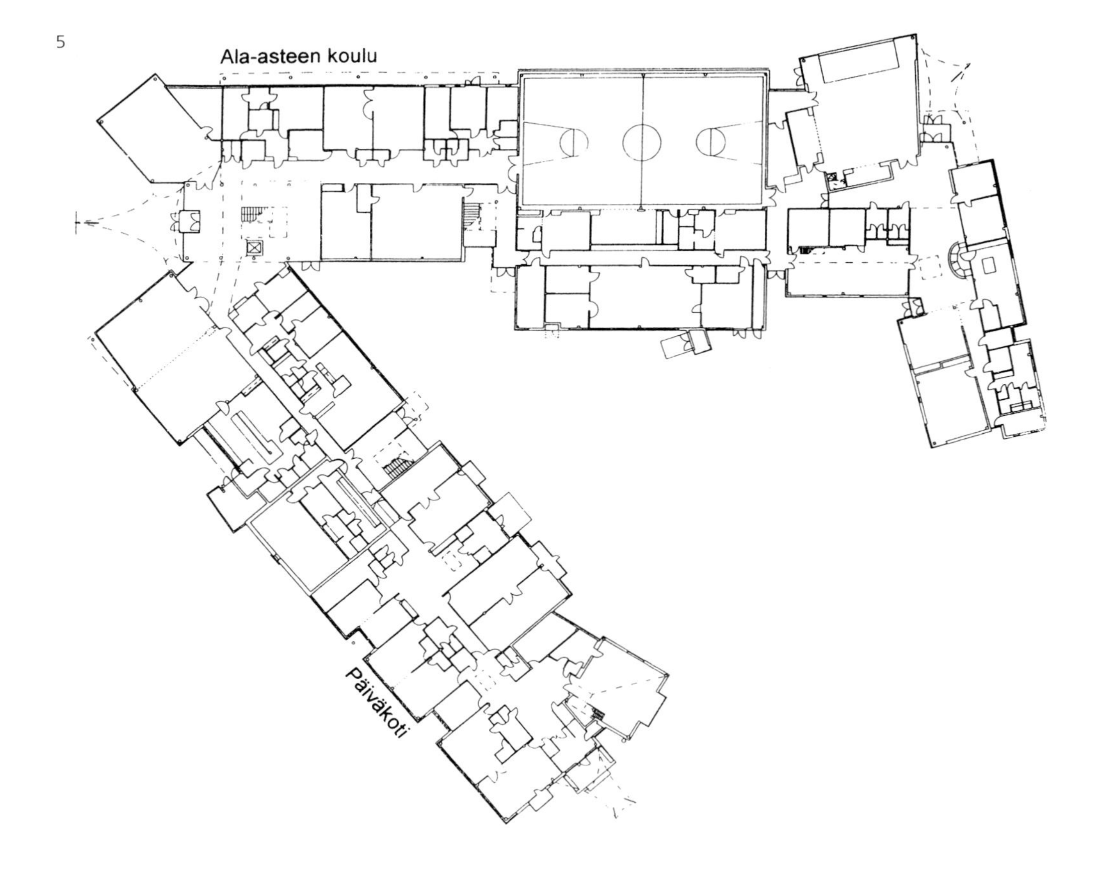

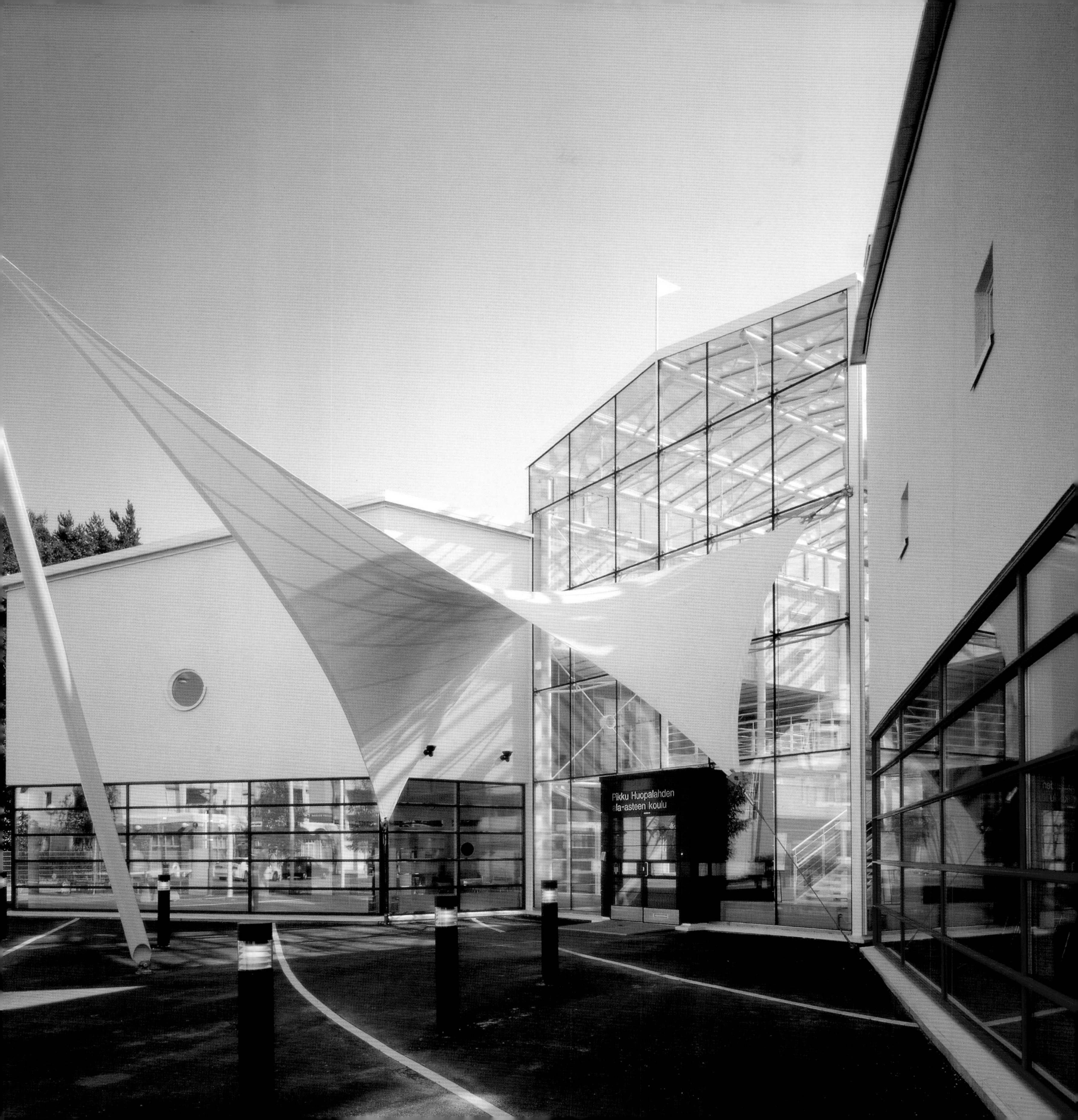
Pikku Huopalahden
ala-asteen koulu

7

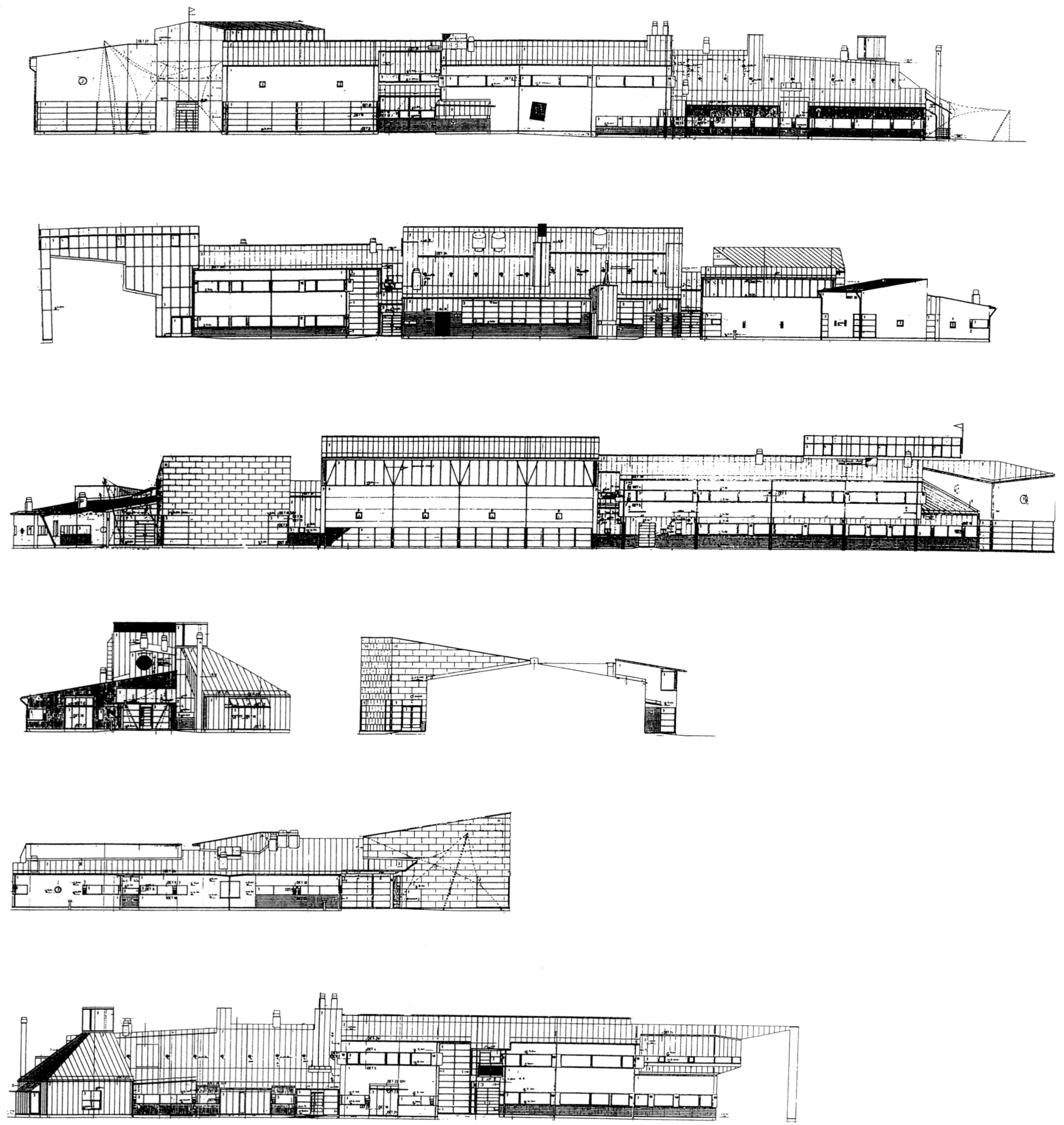

8

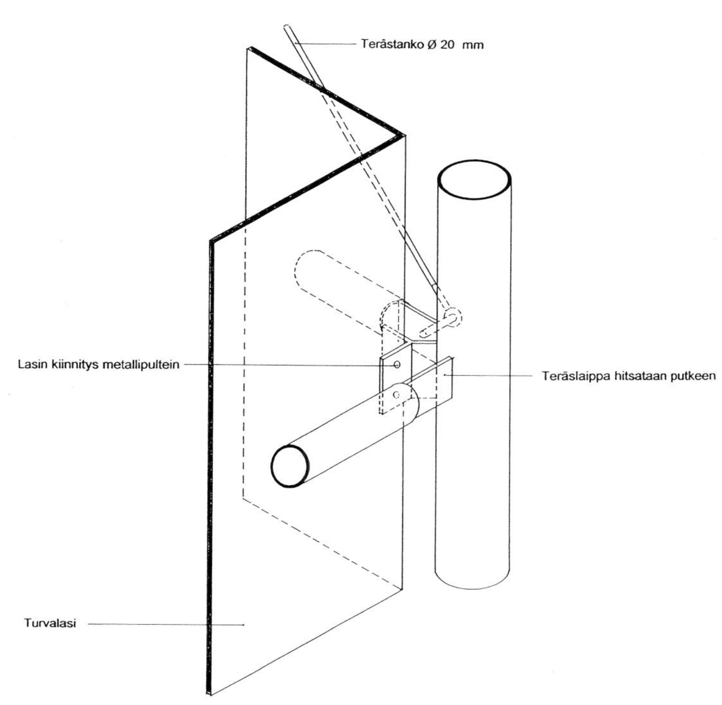

7 立面
8 电梯井细节
9 青年活动中心天篷
10 街道立面

7 Facades
8 Elevator shaft detail
9 Youth centre canopy
10 Street facade

9

10

11 剖面A–A，大厅
12 剖面C–C，日托中心
13 剖面B–B，餐厅
14 入口大厅草图
15 入口大厅内景
16 入口大厅
17 透视
18 日托中心庭院
19 透视
20 日托中心庭院

11 Section A-A, lobby
12 Section C-C, day care centre
13 Section B-B, restaurant
14 Entrance hall sketch
15 View to the entrance hall
16 Entrance hall
17 Perspective
18 Day care centre courtyard
19 Perspective
20 Day care centre courtyard

11

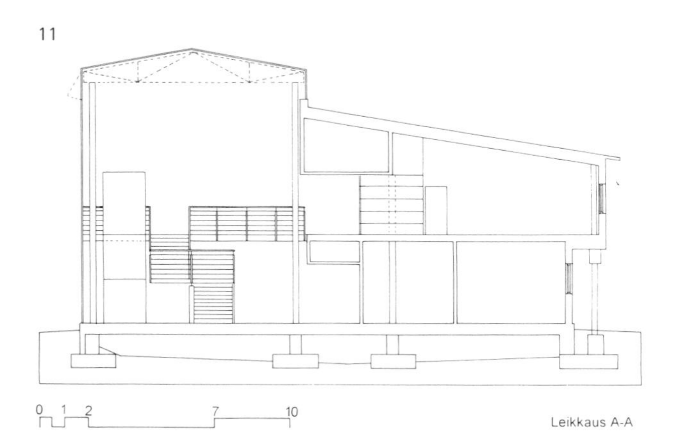

12

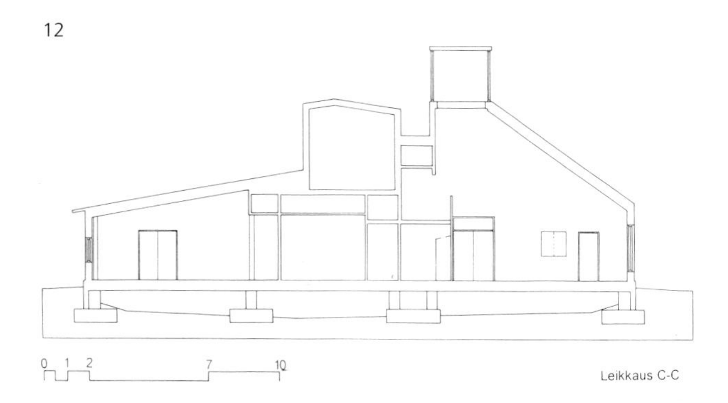

13

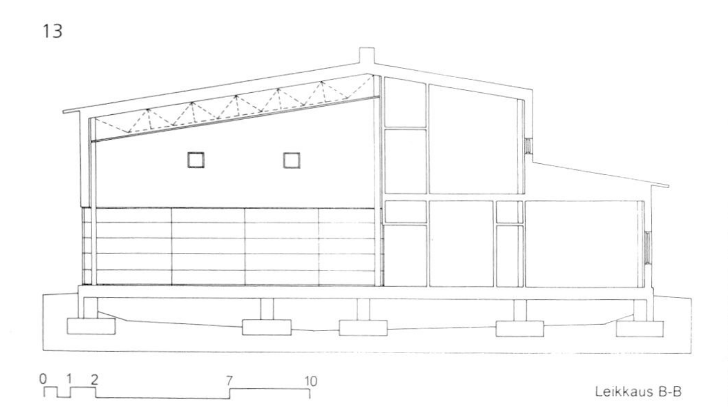

14

15

17

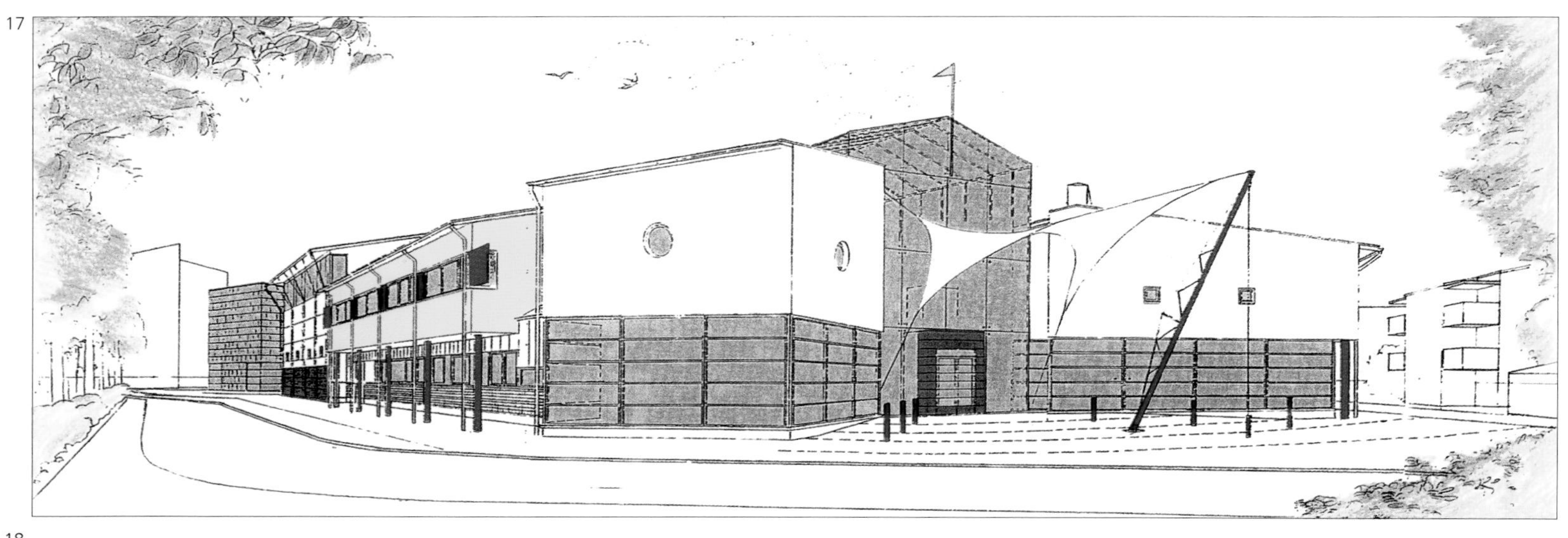

18

19

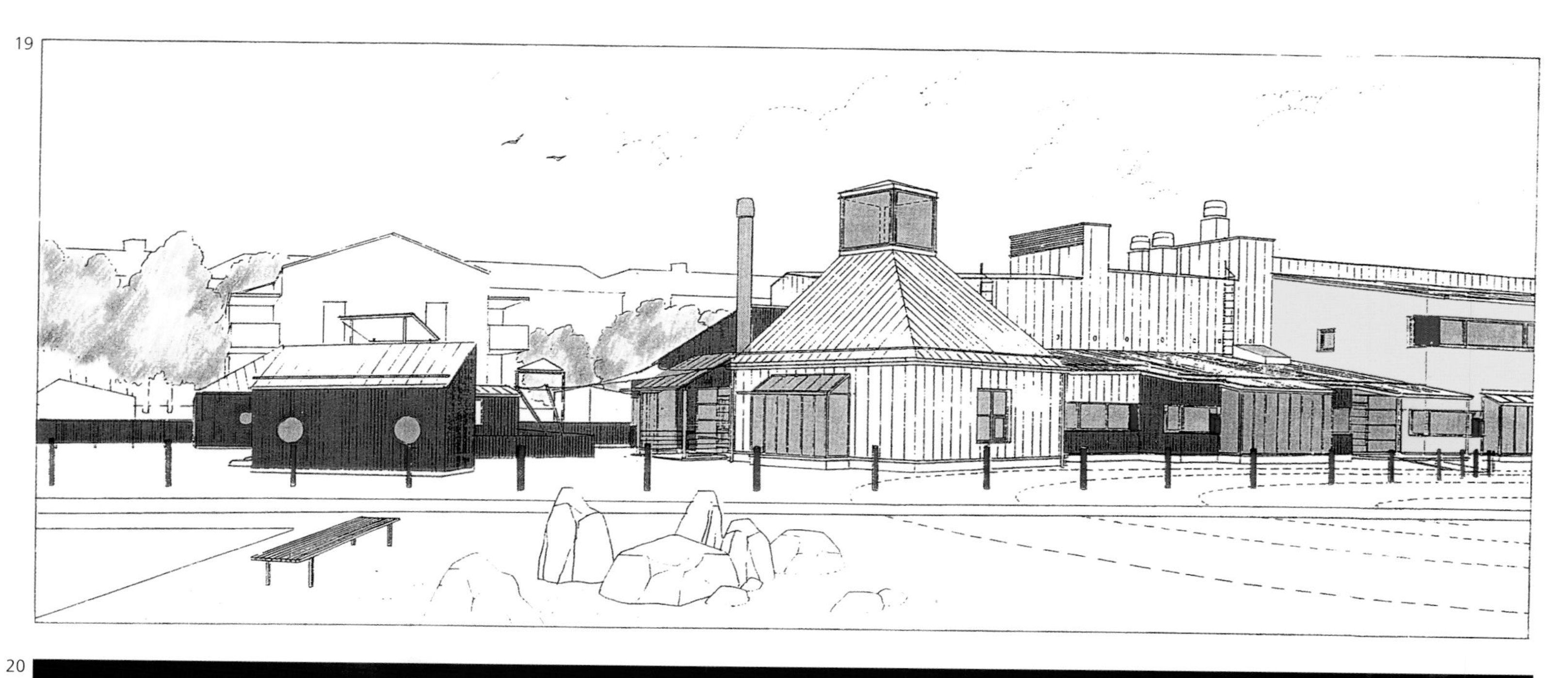

20

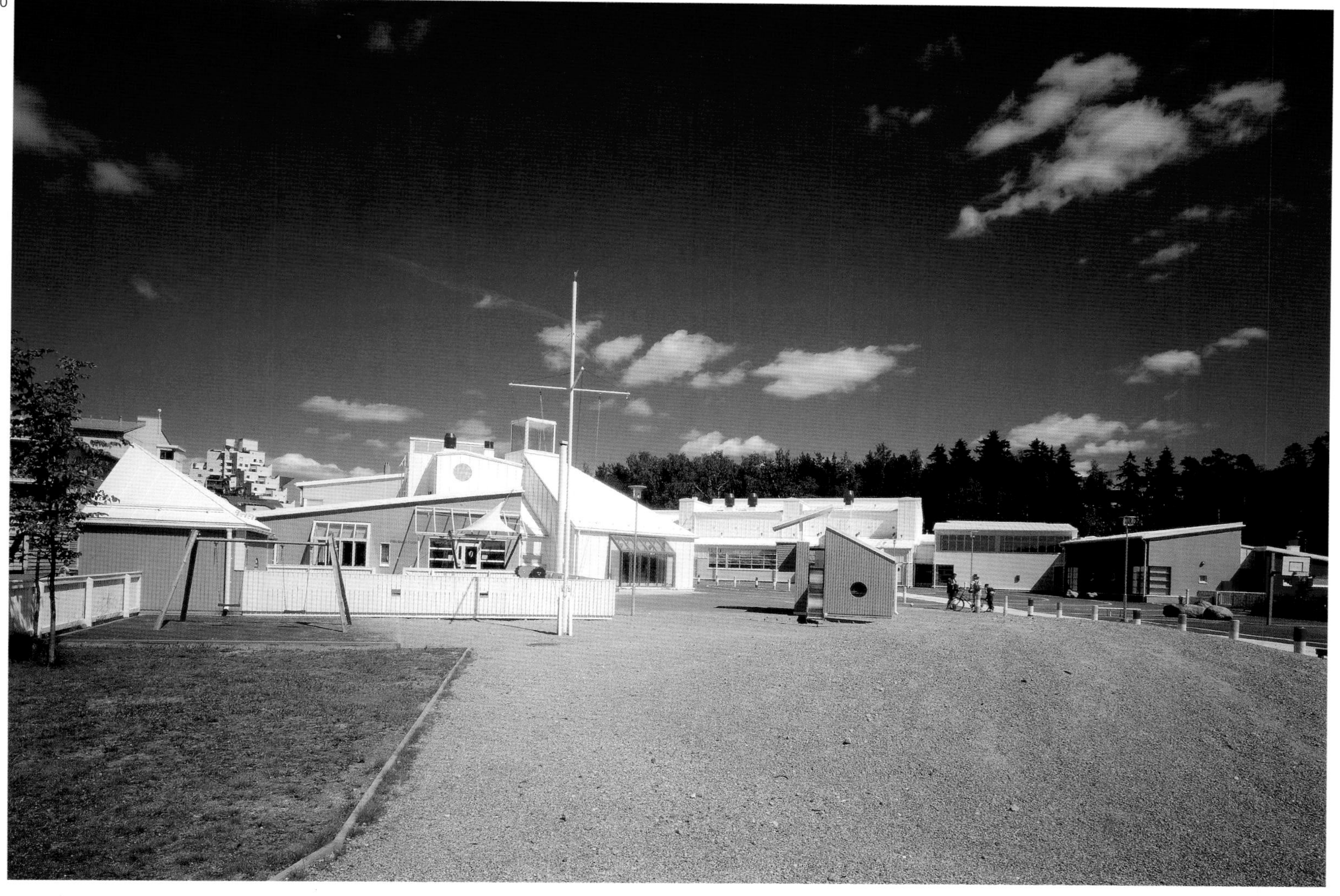

21　学校操场
22　玻璃墙细部
23　日托中心入口天篷

21　Schoolyard
22　Glass wall details
23　Day care centre entrance canopy

21

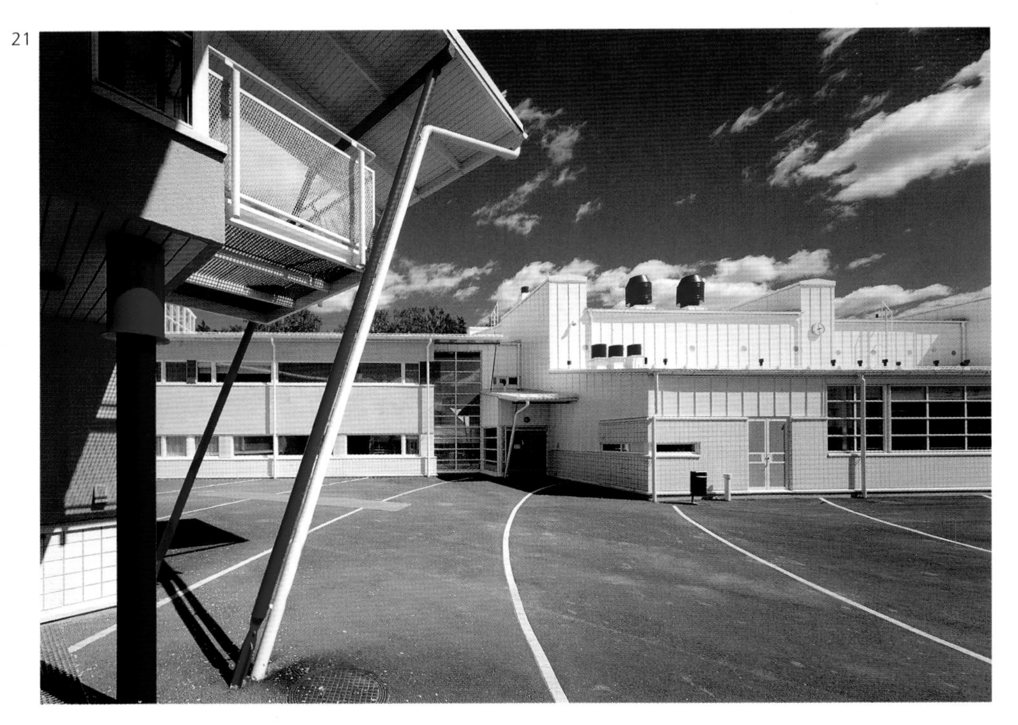

22

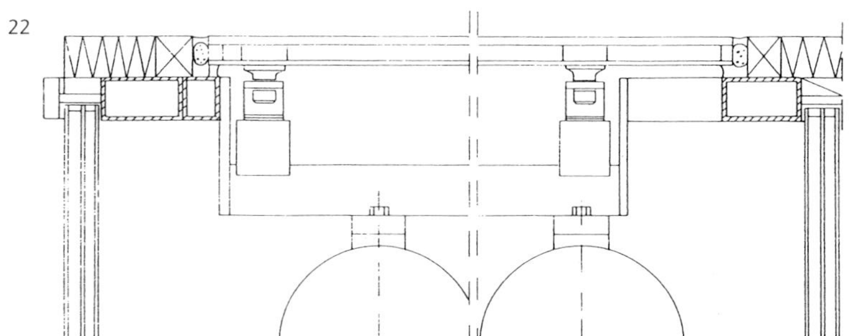

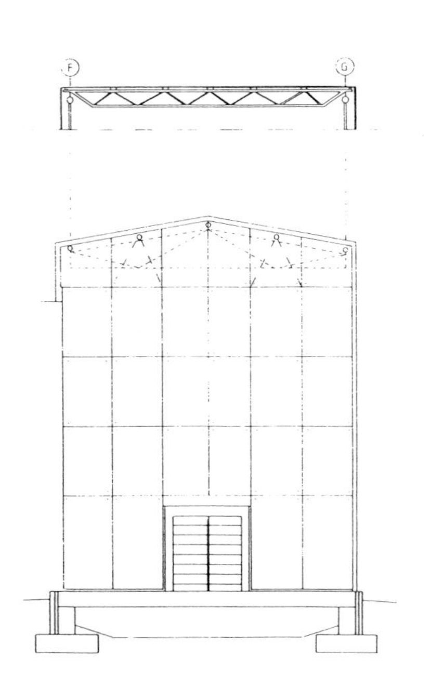

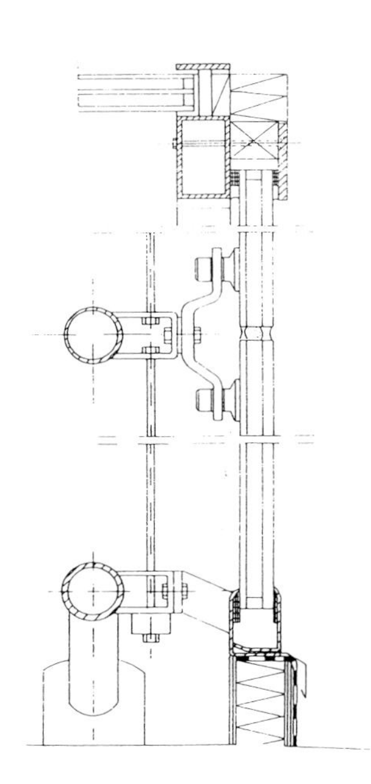

23

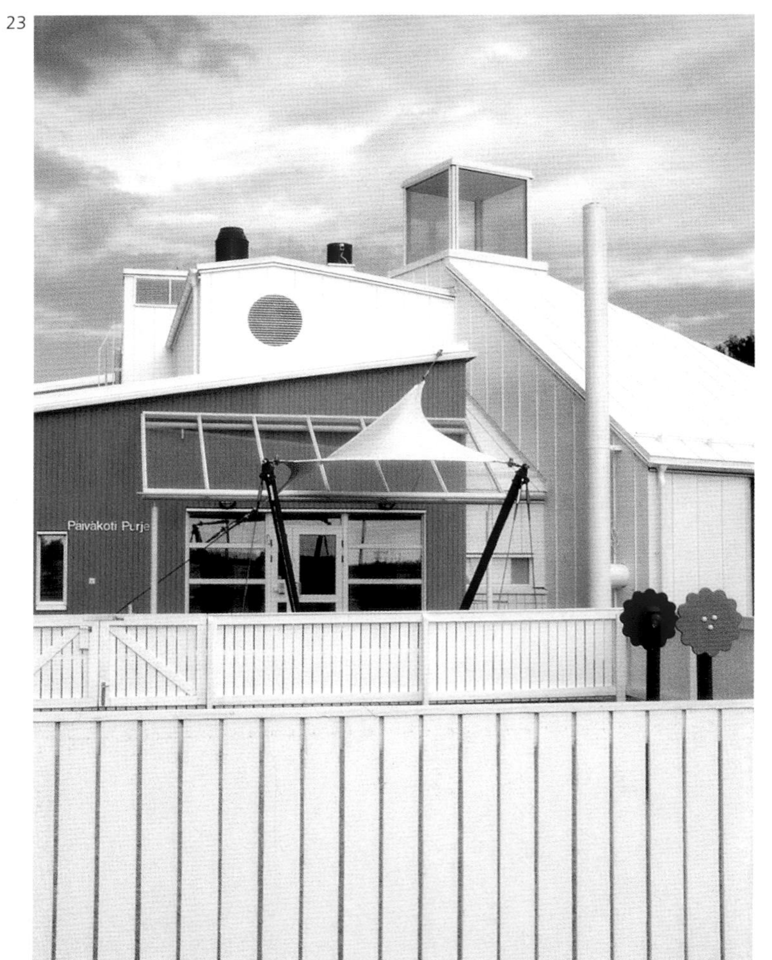

沃萨里地铁站
赫尔辛基 1998

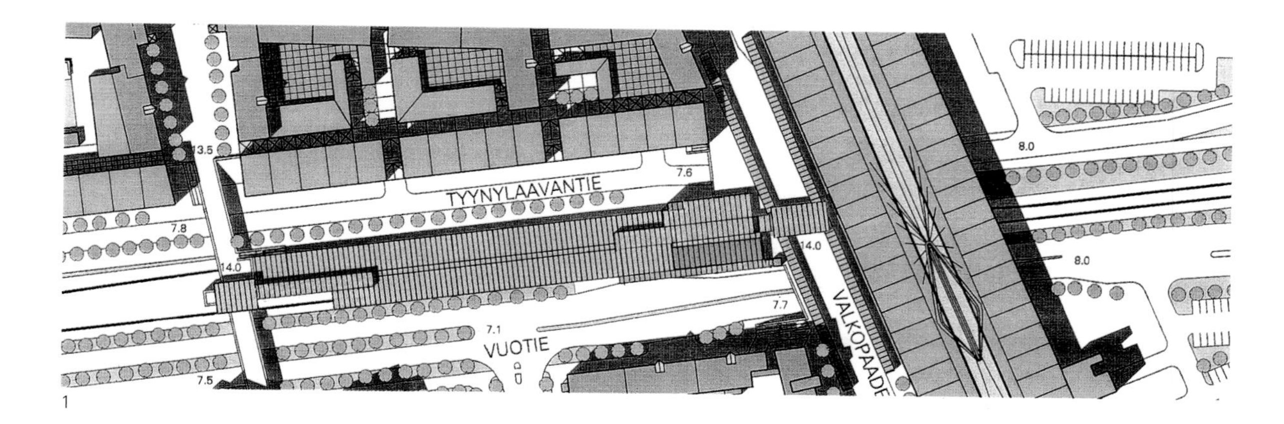

1

2

Vuosaari 地铁线上的车站代表了赫尔辛基新一代的地铁站的建筑水平。运输和交通建筑中的开放式设计反映了我们这个时代民主的建筑方式。这个车站同样也象征了公众运输所具有的吸引力、它的友善和它的安全素质。Vuosaari 地铁站大部分位于地上，空间宽敞光线明亮。站台尽可能开放，在视觉上与外部环境衔接起来。位于售票大厅之间的站台区域上空以玻璃为屋顶。

在这个车站的设计中，关键的问题是提高乘客的安全性和舒适性。在建筑上，强调了这个车站作为这个地区交通枢纽中的一个地标。车站的立面是透明的，给人一种开阔的感觉。

车站最主要的建筑主题是它的玻璃屋顶。站台区域是非对称的。在站台层，车站的外墙由金属网构件做成。

1 总平面
2 西入口大厅

1 Site plan
2 Western entrance hall

Vuosaari Metro Station Helsinki 1998

The stations on Vuosaari metro line represent a new generation architecture of metro stations in Helsinki. The open architecture of transport and traffic reflects the democratic building approach of our time. The openness can also be seen as the attraction, user-friendliness and safety of public transport.

Vuosaari metro station is mostly overground with special attention paid to spaciousness and light. The platforms are visually connected with the environment as openly as possible. The platform area between the ticket halls of the station is covered with glass.

In the design of the station the main focus has been on increasing the satisfaction and safety of the passengers. Architecturally the station has been accentuated as a landmark of traffic in the area. The station facades are transparent, producing a feeling of openness.

The main architectural theme of the station is based on glass roofs. The platform area is asymmetrical. At the platform level the outside walls of the station are made of mesh plate cassettes.

3 汽车终点站玻璃天棚
4 主入口

3 Bus terminal glass canopies
4 Main entrance

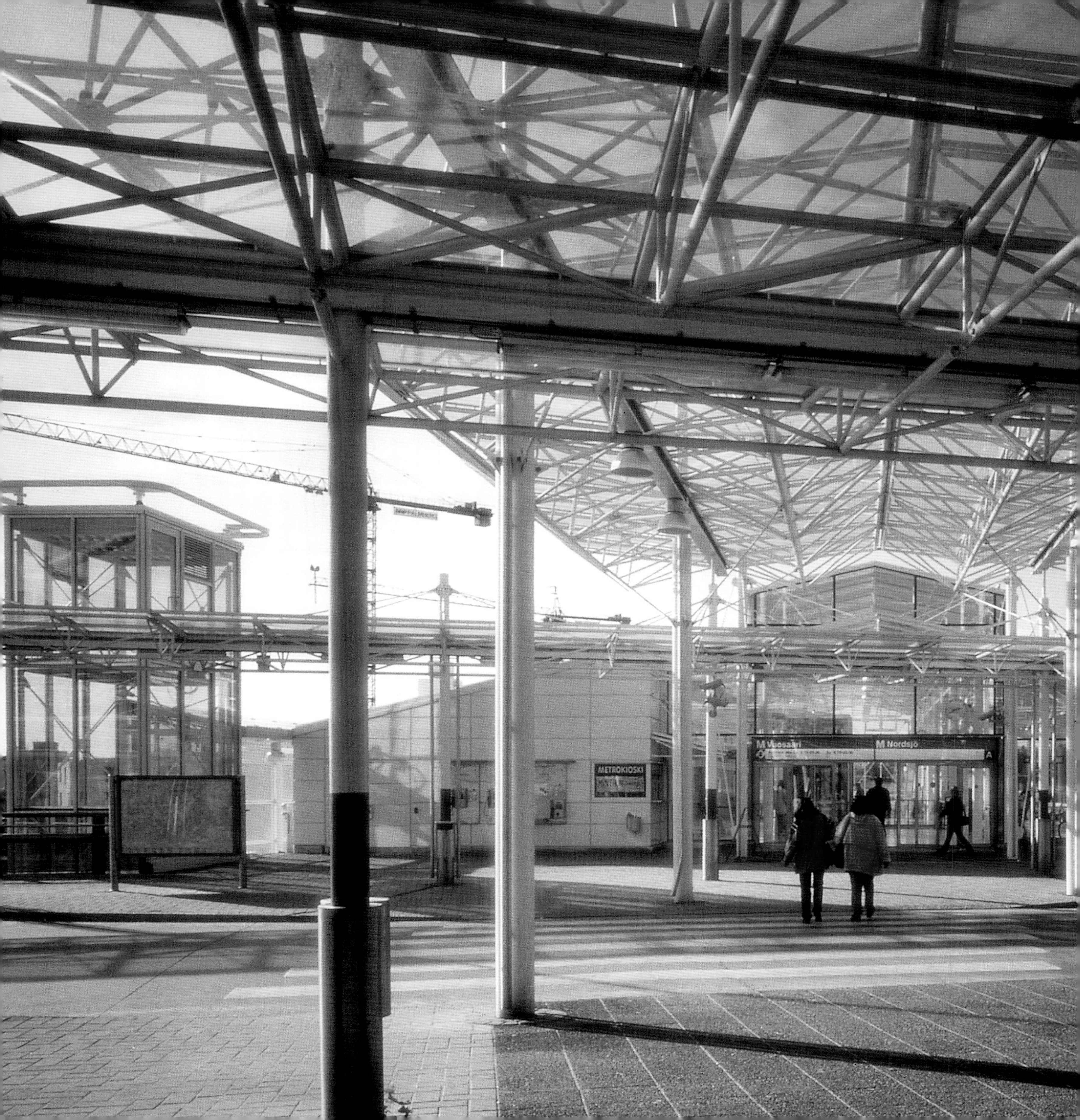
METROKIOSKI
M Vuosaari
M Nordsjö
A

5

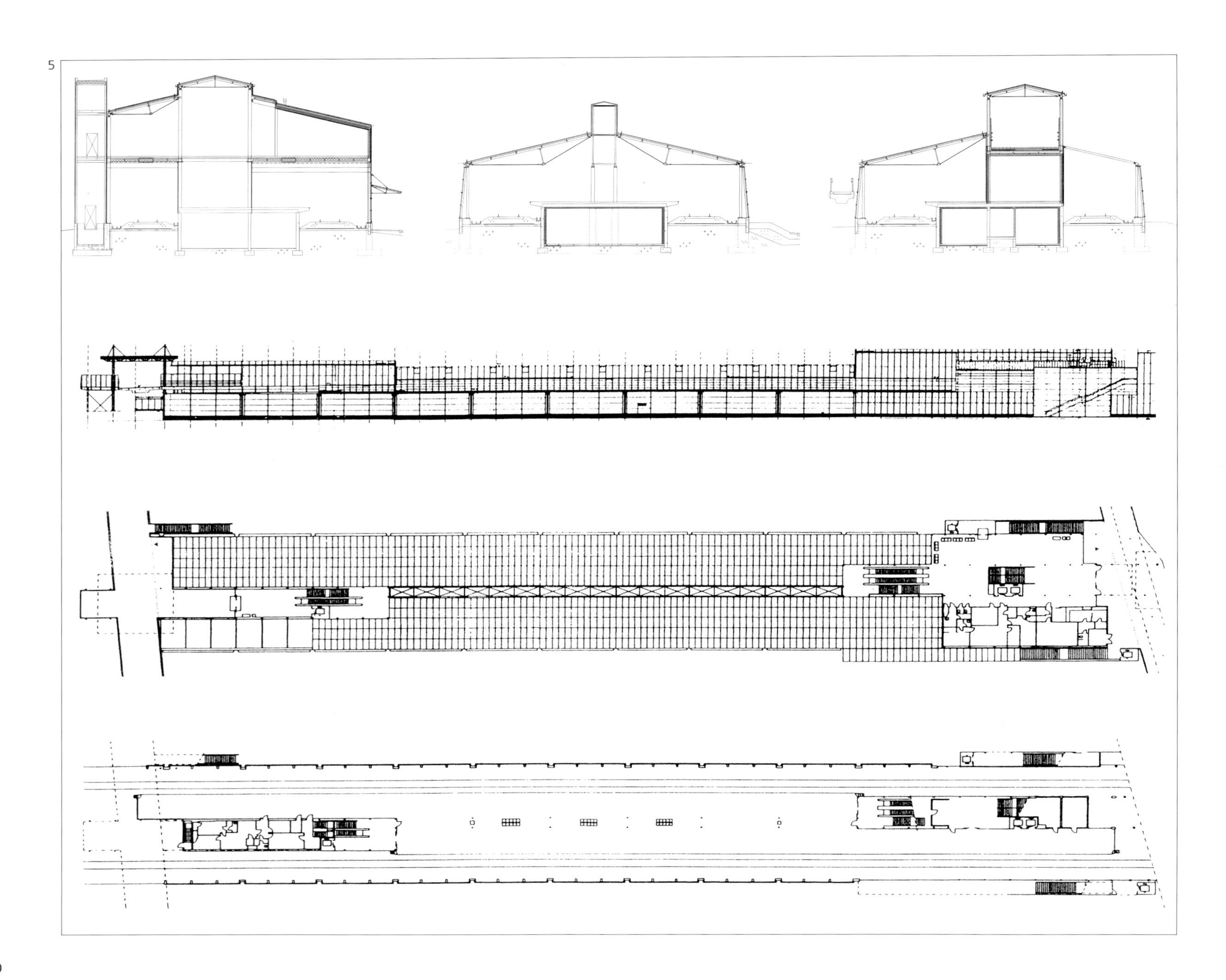

5 剖面、立面和平面
6 外观
7 入口大厅
8 候车大厅
9 候车平台大厅中Jussi Niva创作的艺术品
10 候车大厅

5 Sections, facade and floor plans
6 Exterior view
7 Entrance hall
8 Platform hall
9 Platform hall art work by Jussi Niva
10 Platform hall

6

7

8

9

Vuosaari
Nordsjö

11

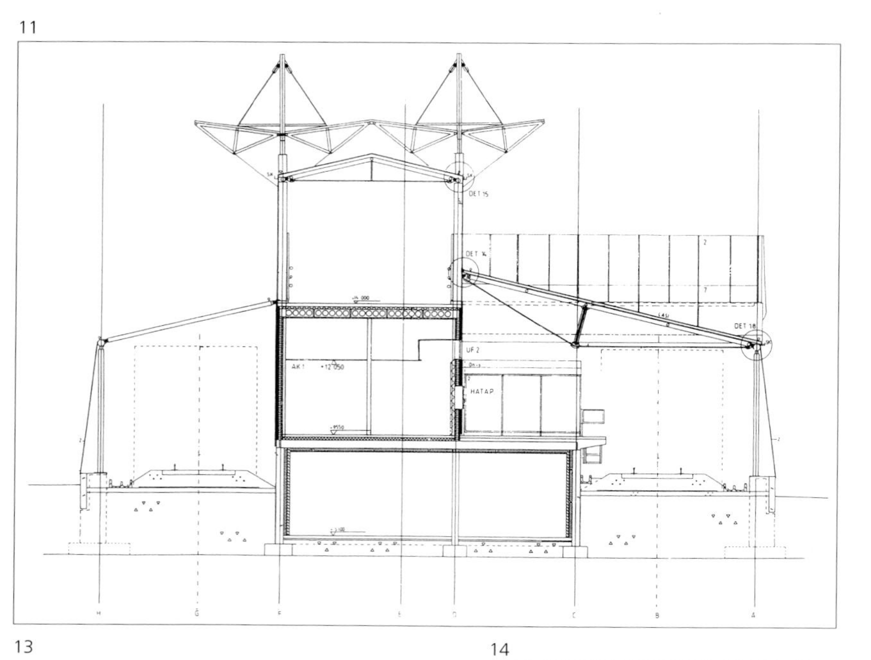

12

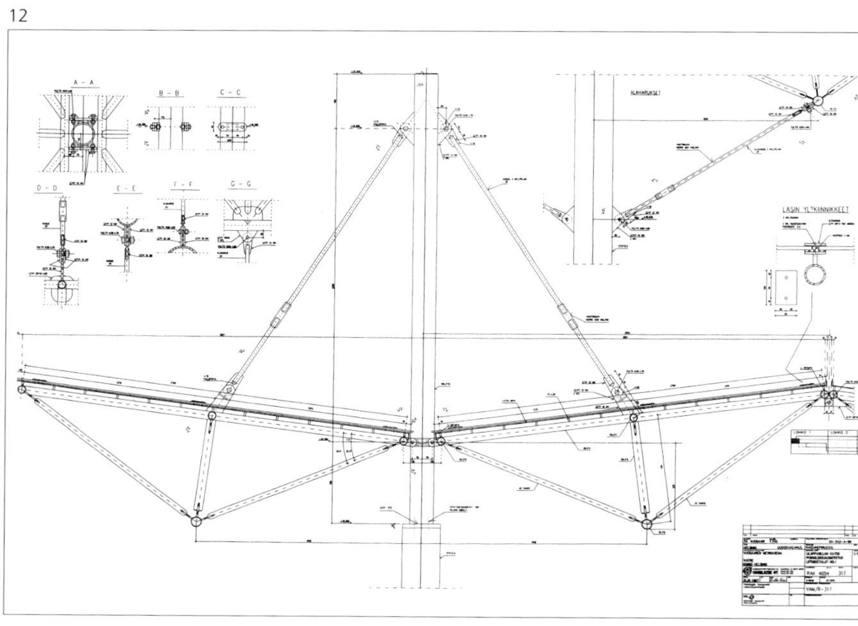

13

14

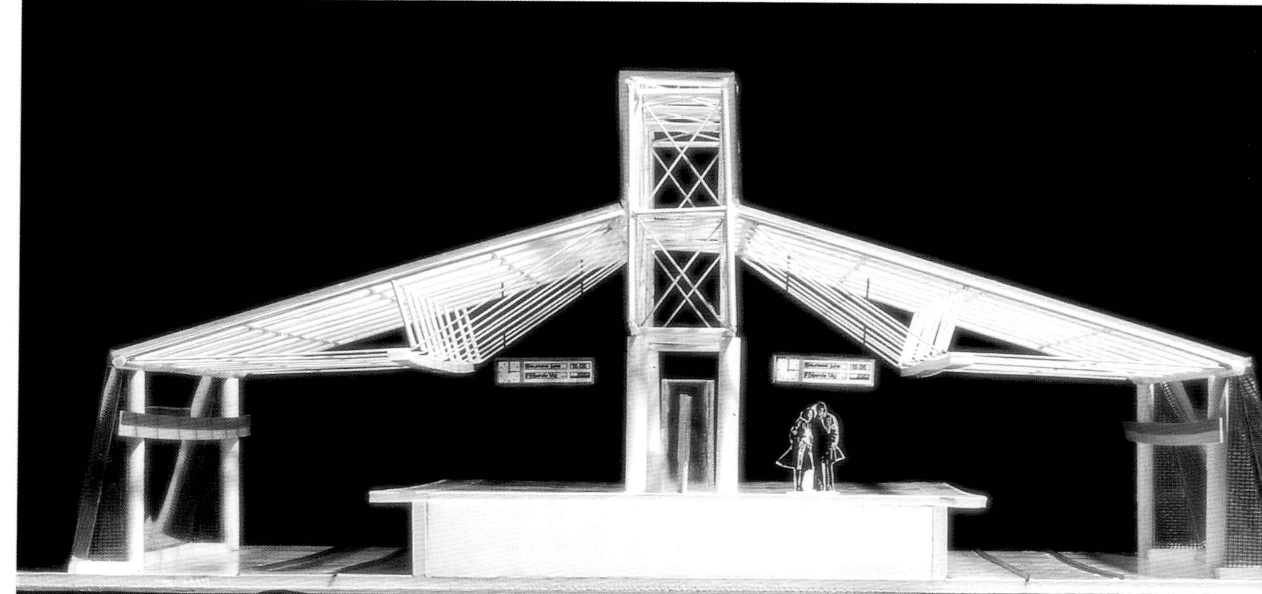

11 剖面
12 雨棚和细部
13 公共汽车站雨棚
14 模型照片
15 玻璃顶细部
16 模型照片

11 Section
12 Canopy and details
13 Bus terminal canopy
14 Model photo
15 Glass roof details
16 Model photo

15

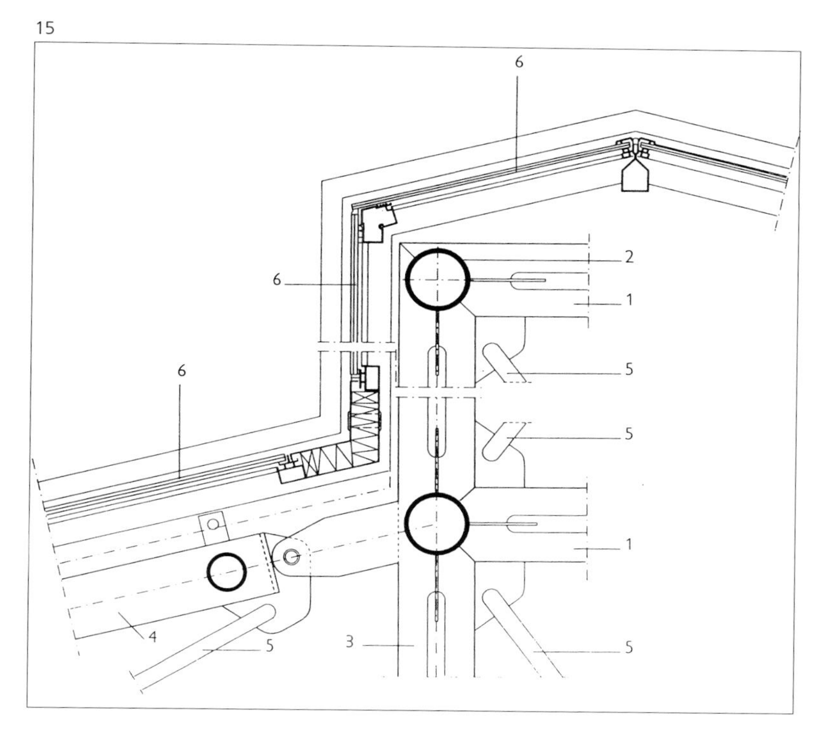

1. 圆形空调剖面Ø 273 × 12.5mm
2. 圆形空调剖面Ø 219 × 12.5mm
3. 圆形空调剖面Ø 273 × 12.5mm
4. 圆形空调剖面Ø 139 × 10mm
5. 钢杆Ø 20mm
6. 由单层白玻和夹胶玻璃组成的隔热玻璃

1. circular hollow section, Ø 273x12.5 mm
2. circular hollow section, Ø 219x12.5 mm
3. circular hollow section, Ø 273x12.5 mm
4. circular hollow section, Ø 139x10 mm
5. steel rod, Ø 20 mm
6. insulating glass of single float glass and laminated glass

16

珀里火车站
隧道和站台顶棚
1998

隧道是一个正在设计阶段的交通中心的一部分，这个交通中心将各种不同的公共交通形式（包括航空、铁路和公交巴士）结合在一起。

2号主干线下的地下穿越道与隧道入口的设计相适应。Pori 的隧道入口是街道节奏的延续。明确的方向感和清晰的空间性是设计最为重要的出发点。楼梯和电梯边的采光口以及隧道中部的天窗增加了隧道内的光感。隧道的端部是钢和玻璃结构的顶棚，强调出隧道入口的形象。隧道就像是一座悬挑的平板桥。由于隧道的墙体是玻璃，因此光线可以通过它们成倍地反射出来，从而给人一个虚幻空间的感觉。霓虹灯的图案更增添了这种气氛。

地面上的楼梯井和电梯轿厢是钢化镀膜玻璃和简单的钢结构建造的。站台顶棚由钢框架支撑，柱子是圆形的管子。顶棚是由圆管和杆件组成的空间结构。顶棚结构从柱子的顶部挑出来。站台屋面是10mm厚钢化玻璃，节点为黏结。透明的站台顶棚创造了一个开阔明亮的空间，而开阔且透明的建筑则增加了乘客的安全性。

1 总平面
2 透视
3 通向候车平台的楼梯

1 Site plan
2 Perspective
3 Stairs up to the platforms

1

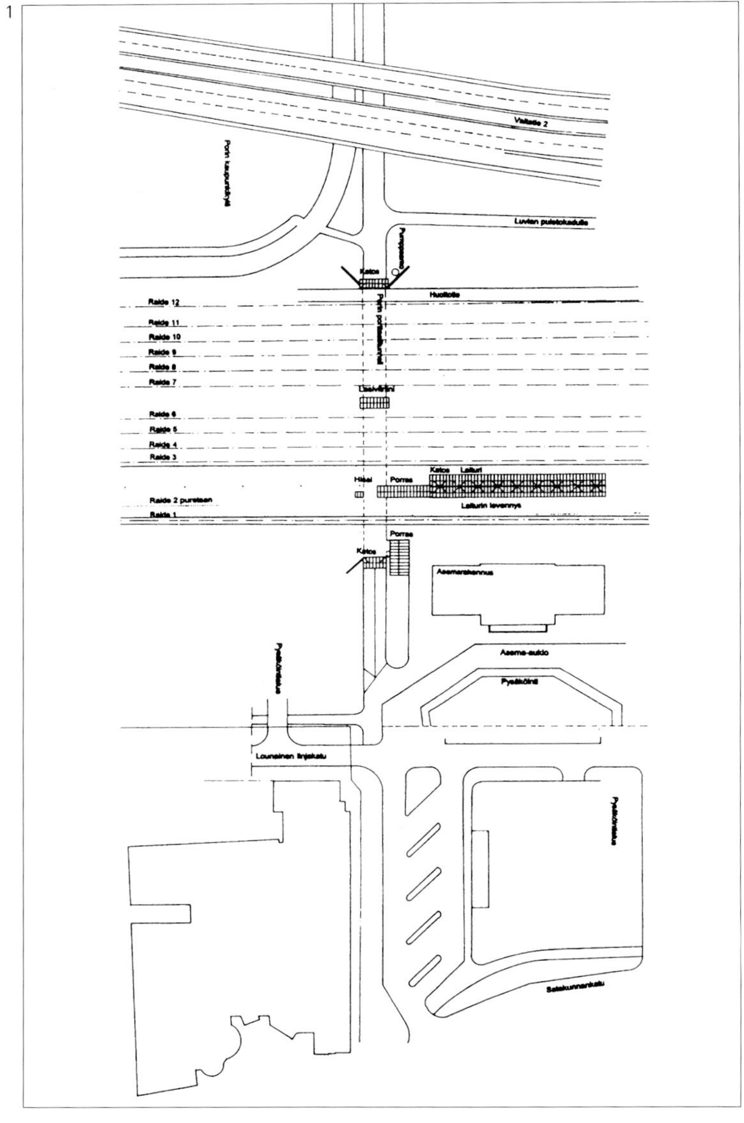

2

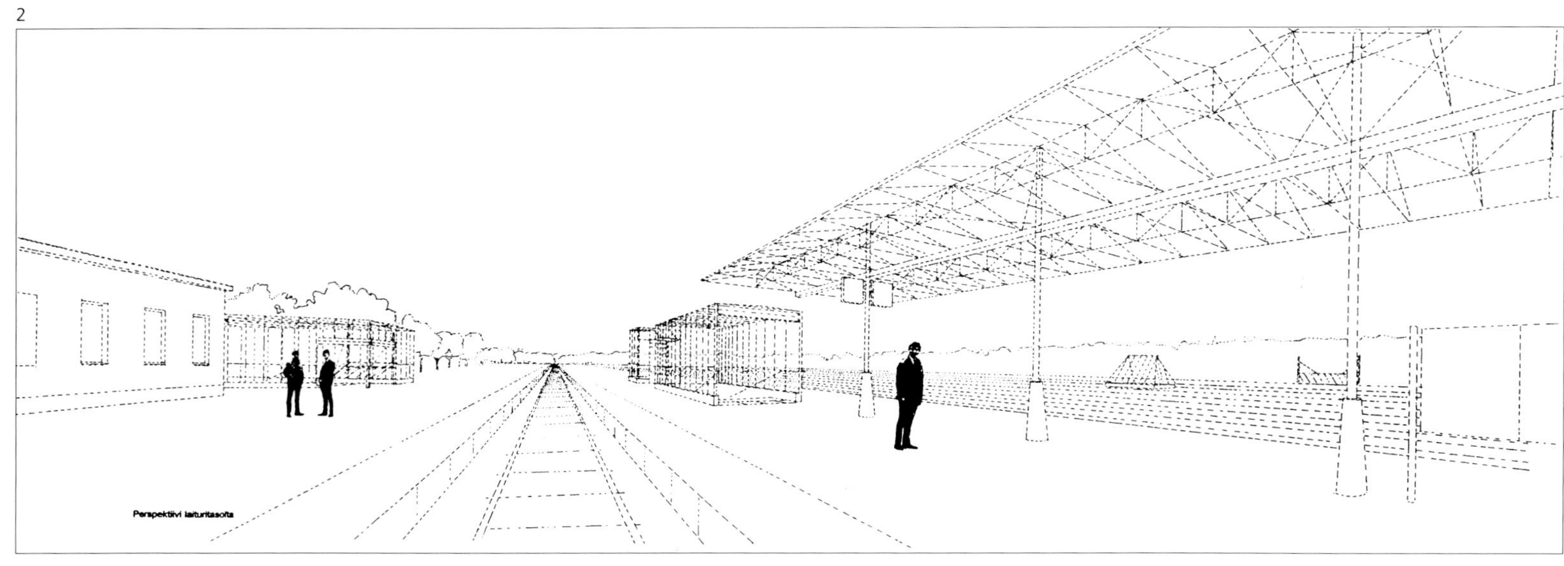

Raide
Spar
Track
3
3

Pori Railway Station Tunnel and Platform Shelters 1998

4 人行隧道入口
5 候车平台平面
6 隧道层平面和剖面
7 钢和玻璃雨棚

4 Pedestrian tunnel entrance
5 Platform level floor plan
6 Tunnel level floor plan and section
7 Steel and glass canopy

4

The tunnel is part of a travel centre which is still at a design stage and will combine the different forms of public transport (air, rail and bus traffic).
The underpass on trunk road 2 has been adapted to the portal tunnel design.
The portal tunnel in Pori is a continuation of street pace, with orientation and spatial clarity being the most important starting points in the design work. Light openings by the stairs and lifts as well as the glass lantern in the midtunnel increase the impression of light in the tunnel. The tunnel ends are covered with steel-glass shelters which emphasize the portal tunnel image.
The tunnel has been built as a cantilever slab bridge. As the tunnel walls are of glass, the light reflected through them is multiplied, creating in illusory space impression. The atmosphere is further emphasized by the symbol patterns constructed of neon lights.
The staricase shelters and the lift shaft overground are of annealed and laminated glass with simple steel structures. The platform shelter is built on a steel frame, and the columns are round pipe profiles. The frame of the roof structure is realized as a space frame structure, built of round pipes and rods.
The roof structure is suspended from the columns on the top side. The platform shelter roofing is 10 mm annealed glass with glued joints. The transparency of the platform shelter creates an open and light entity, and the open and transparent architecture increases the safety of the passengers.

5

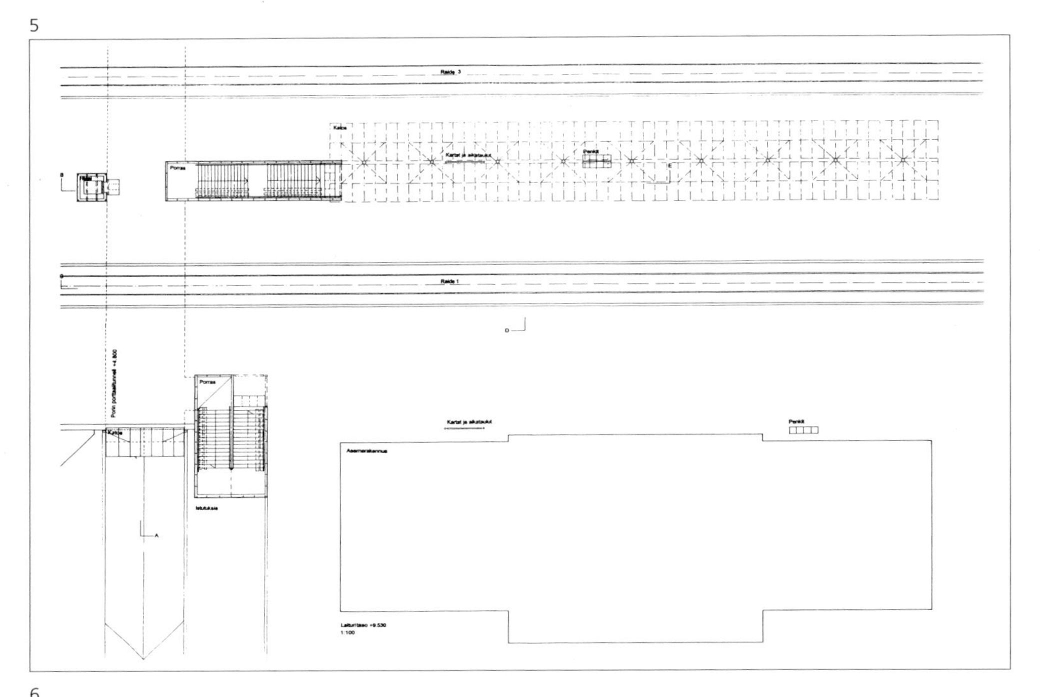

6

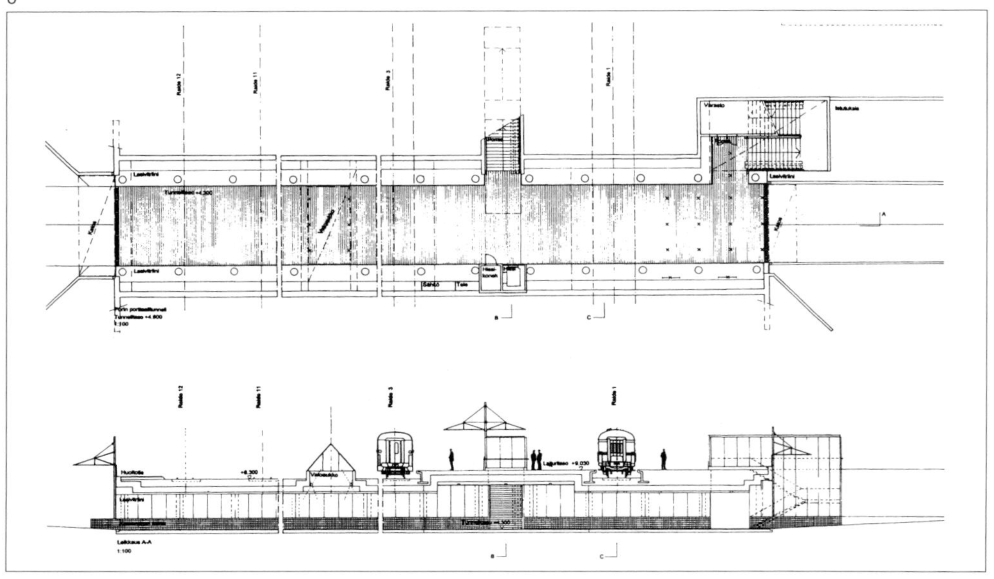

8

9

10

11

12

13

8　细部
9　楼梯
10　候车平台顶棚
11　楼梯入口
12　从隧道看
13　细部
14　候车月台遮棚

8 Detail
9 Staircase
10 Platform shelter
11 Staircase entrance
12 View from the tunnel
13 Detail
14 Platform shelter

14

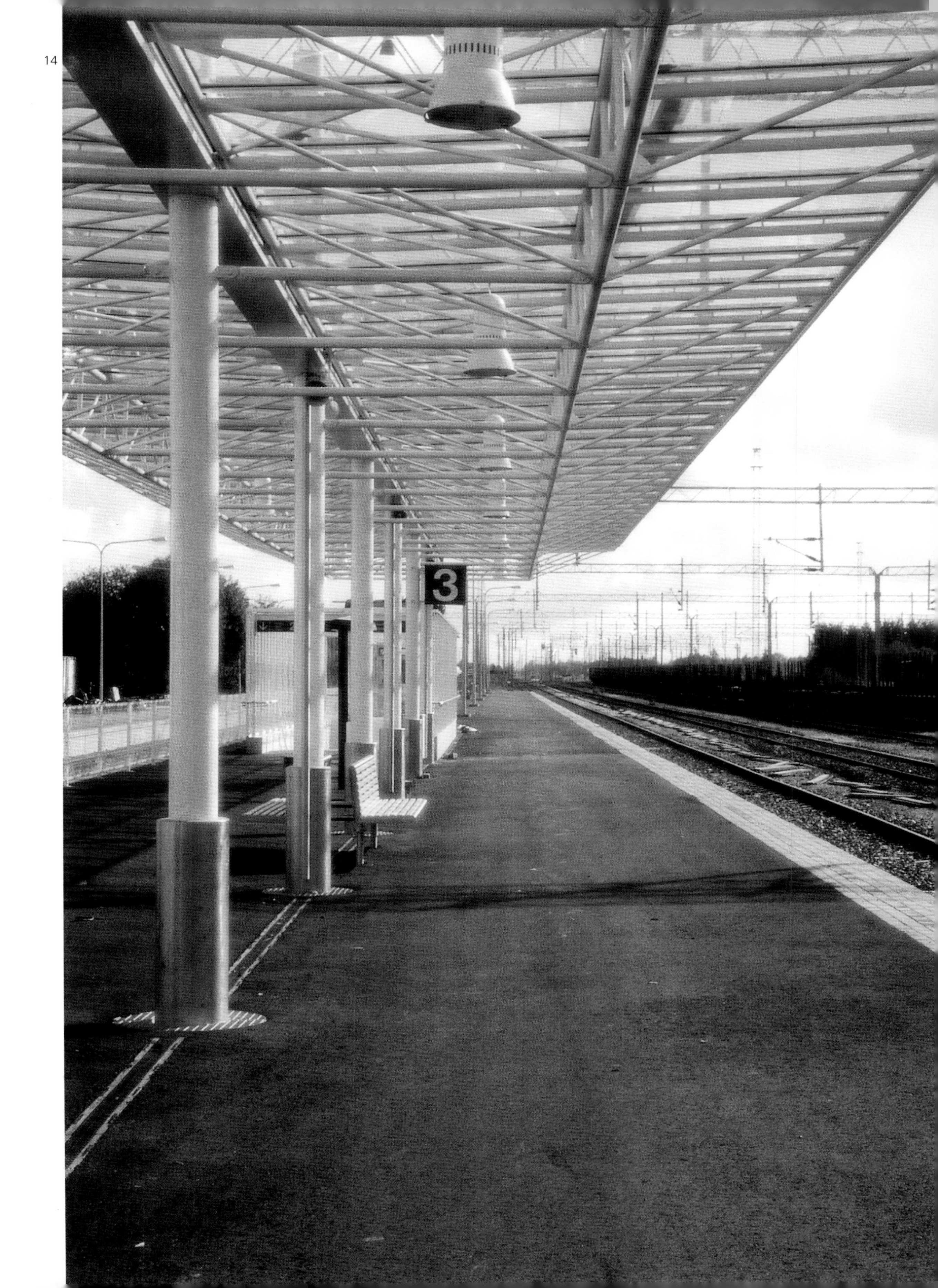

VR

1

科尔索火车站，万塔 2000

Korso Railway Station, Vantaa 2000

科尔索车站的布置与连接Rekola和科尔索之间新的辅助轨道的铺设相关联。因为这样的原因，科尔索车站从原来的老车站向北移了个位置。

对科尔索广场下面地下通道进行了拓宽，同时建造了一条新的封闭楼梯，将地下隧道与中心站台相连，这样可以使来往乘客的交通更为方便。西面的轨道是为高速列车预留的。从科尔索广场通往人行和自行车道的楼梯与车站建筑平行，它同样也是加有顶盖的。

科尔索车站的北端建造了一条通向中心站台的跨线桥，在跨线桥的线路上有加顶盖的楼梯和坡道。跨线桥与车站外的人行和自行车道之间，通过轻钢结构建造的楼梯连接起来。西面的楼梯也是加有顶盖的。在位于南部的楼梯顶盖和位于北部的坡道顶盖之间是钢结构的候车平台的屋面，其中一部分是玻璃顶。屋面是支撑在间距为6m的钢柱之上的张拉杆结构。

楼梯和坡道的顶棚也是钢结构。立面是网格板，这样既保证了车站区充分的采光，同时也有效地阻止了不文明的破坏行为。

混凝土墙体外是瓷砖饰面。桥的基座和其它可见的混凝土部位是涂料面层。

The station arrangements in Korso are connected with the construction of the new auxiliary track between Rekola and Korso. In this connection, Korso station is moved from its present location north, where the old station building used to be.
The underpass running under Korso Square is widened and a new, covered staircase built from the tunnel to the central platform, which serves commuter transport. The western tracks are reserved for express trains. The staircase leading from Korso Square to the pedestrian and bicycele road running parallel with the station, is covered.
An underbridge with covered staircases and ramps leadings to the central platform is built at the north end of Korso station. This 'Kotkansiipi underbridge' is connected with the pedestrian and bicycle roads running outside the station by means of light steel staircases. The western staircase is covered in this connection. A platform shelter of steel construction is built between the southern staircase cover and the northern ramp cover. Part of the platform is covered with glass. The frame of the shelter is supported with draw bars to steel columns placed at 6m spacing.
The structures of the staircase and ramp covers are of steel. The facades are made of grid panels to ensure sufficient lighting in the station environment and to minimize attacks of vandalism.
The concrete structures on the walls are covered with ceramic tiles. The concrete surfaces on the bridge bases and other visible surfaces are painted.

1 候车月台遮棚

1 Platform shelter

2 立面
3 隧道层平面
4 候车月台层平面
5 候车月台遮棚与背景中的旧车站建筑

2 Facades
3 Tunnel level floor plan
4 Platform level floor plan
5 Platform shelter and old station building in the background

2

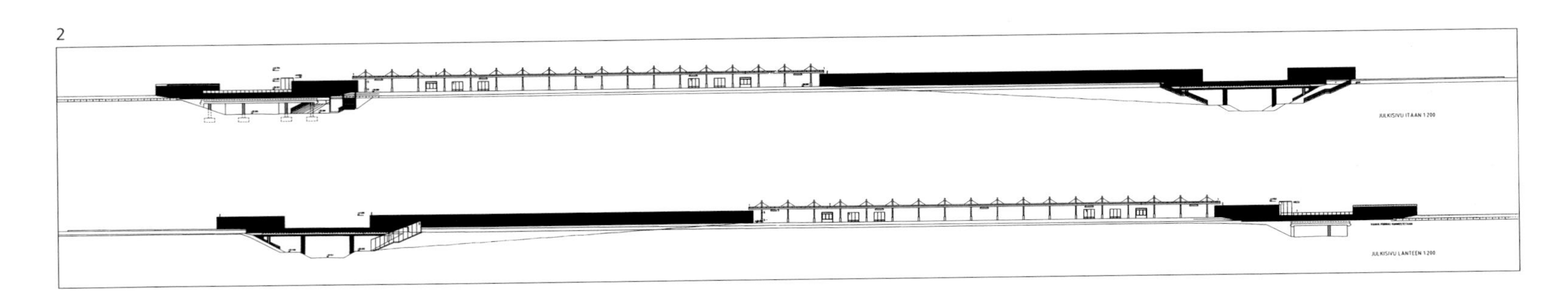

3

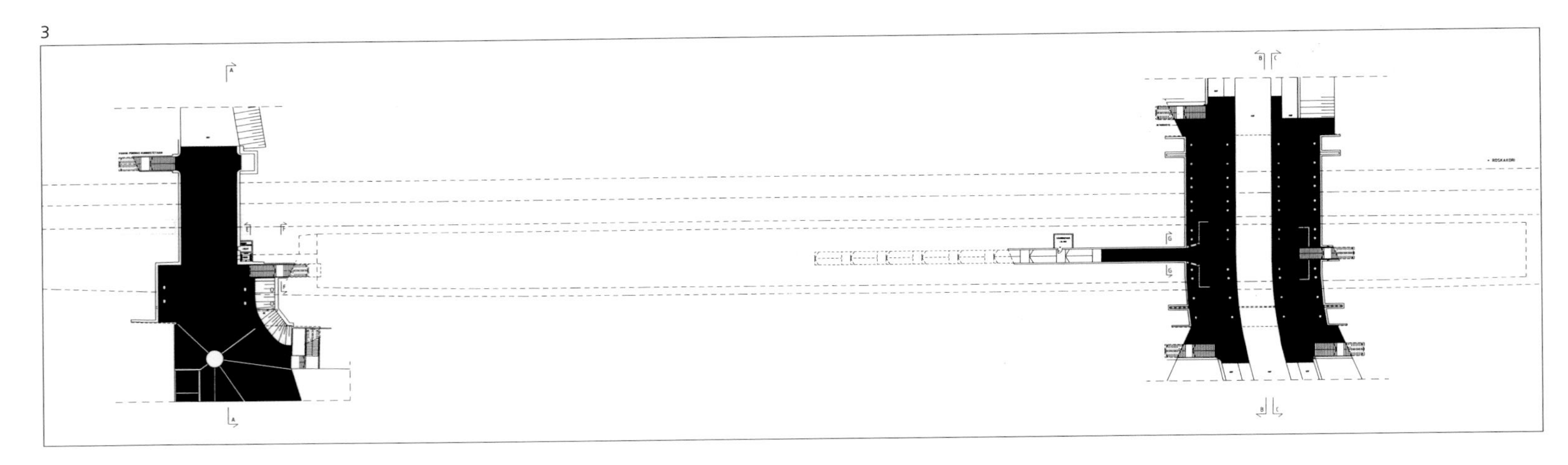

4

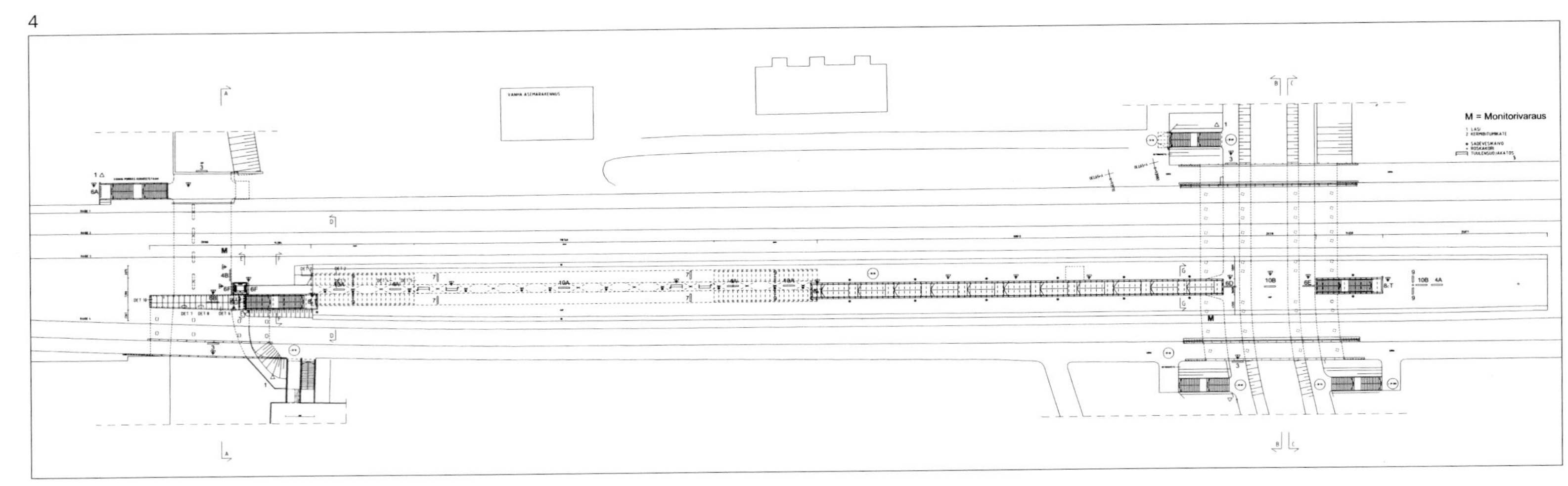

5

6 细部
7 细部
8 细部
9 候车月台遮棚
10 从南面看
11 隧道桥
12 剖面

6 Detail
7 Detail
8 Detail
9 Platform shelter
10 View from the south
11 Tunnel bridge
12 Section

6

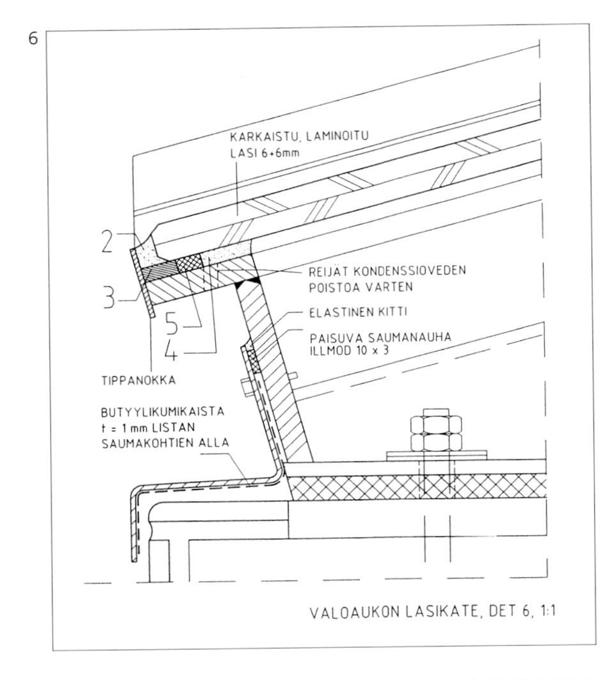

7

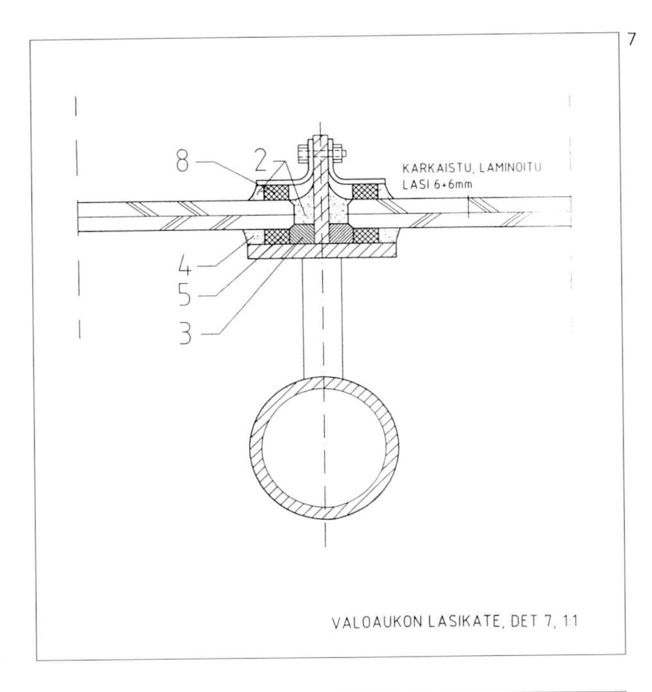

8

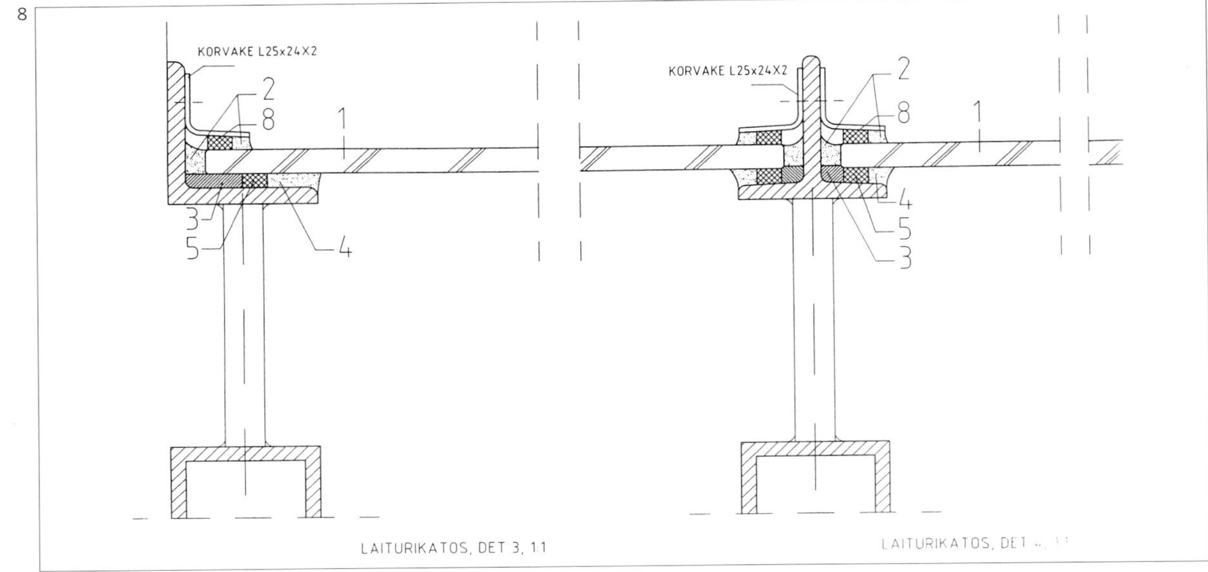

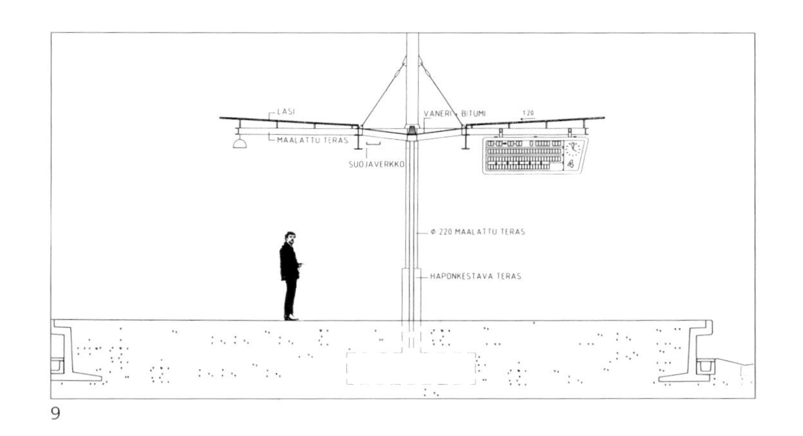
LASI
VANERI + BITUMI
120
MAALATTU TERAS
SUOJAVERKKO
4
Φ 220 MAALATTU TERAS
HAPONKESTAVA TERAS

9

10

11

3
4

12

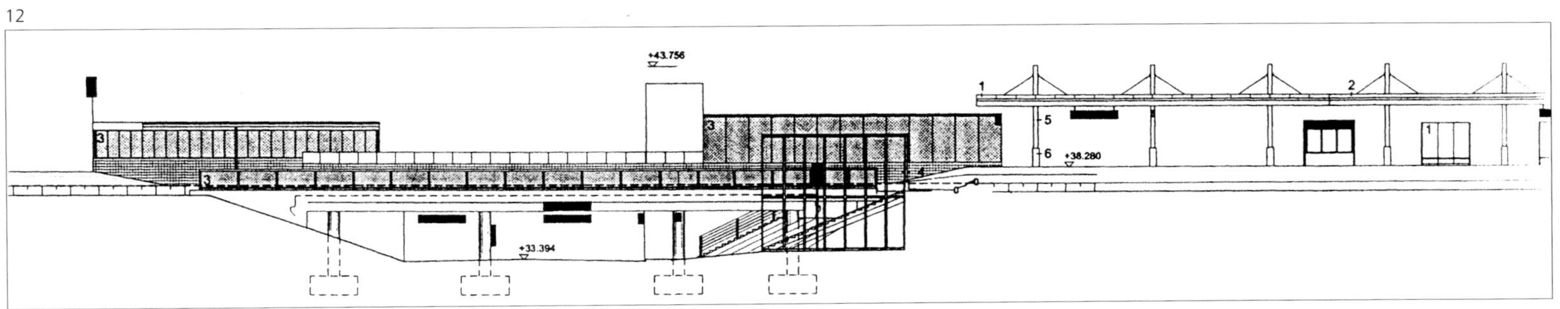
+43.756
1
2
3
5
6
+38.280
+33.394

普梅斯塔里人行天桥，珀里 2001

Pormestari Pedestrian Bridge, Pori 2001

1 总平面
2 剖面
3 平面
4 来自桥上的视点

1 Site plan
2 Section
3 Plan
4 View from the bridge

这座桥的设计在一个获胜方案的基础上发展而来的。

在Pormestarinluoto和Kirjurinluoto之间设计了一座较短的桥。

设计的原则是使得桥的整体形象跟这座城市协调起来。精致的构造使得这个地区看起来富有特点。桥是用钢材建造的。

而桥面则是用木地板铺就。桥的栏杆也全部是木制的。

The design of the bridge is based on a winning entry.
A bridge of light traffic has been planned as a short bridge structure from Pormestarinluoto to Kirjurinluoto. The aim was to make the main expression of the bridge harmonious in the image of the city. The delicate structures give the area its identity.The structures of the bridge are made of steel.
The bridge has a wooden deck. The railing of the bridge is totally of wood as well.

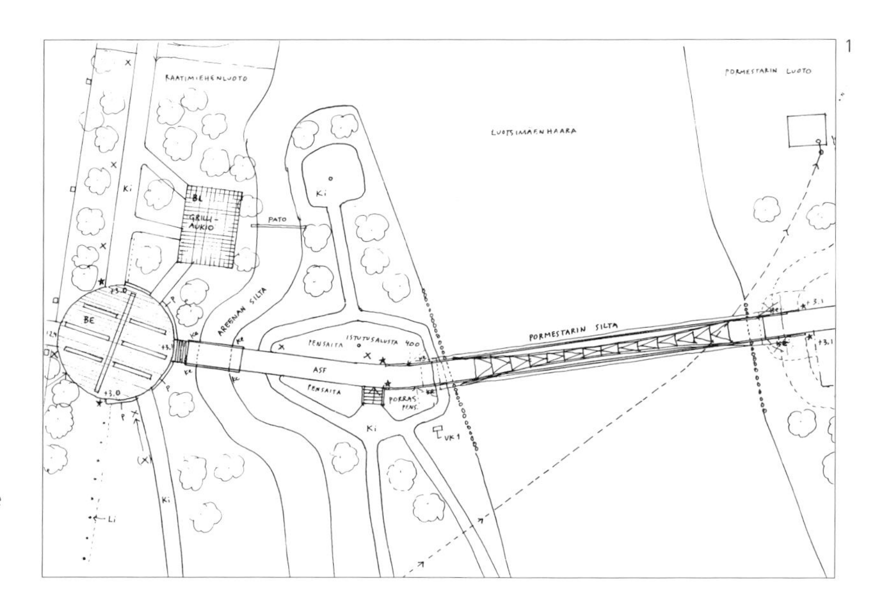

1

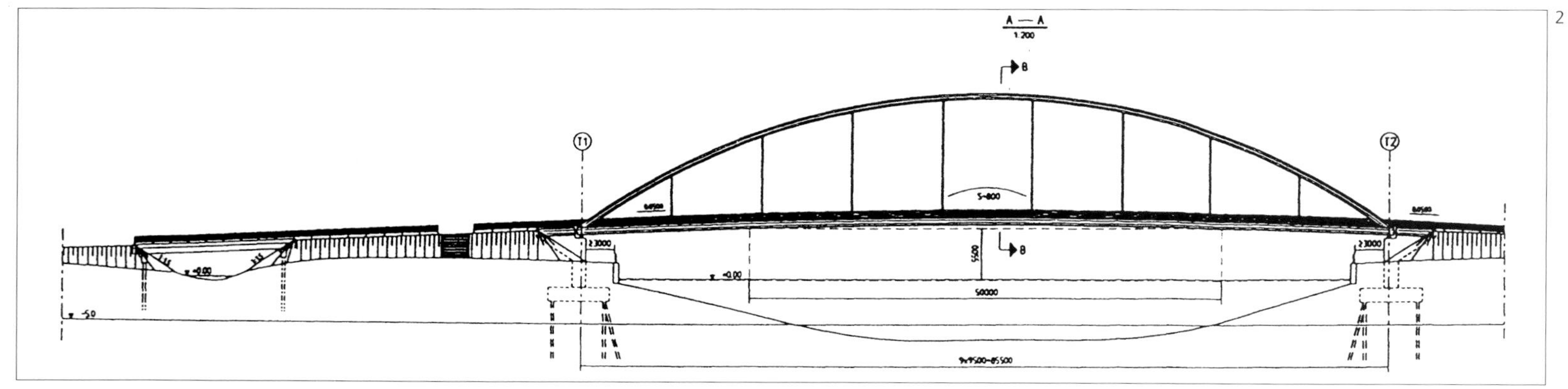

2

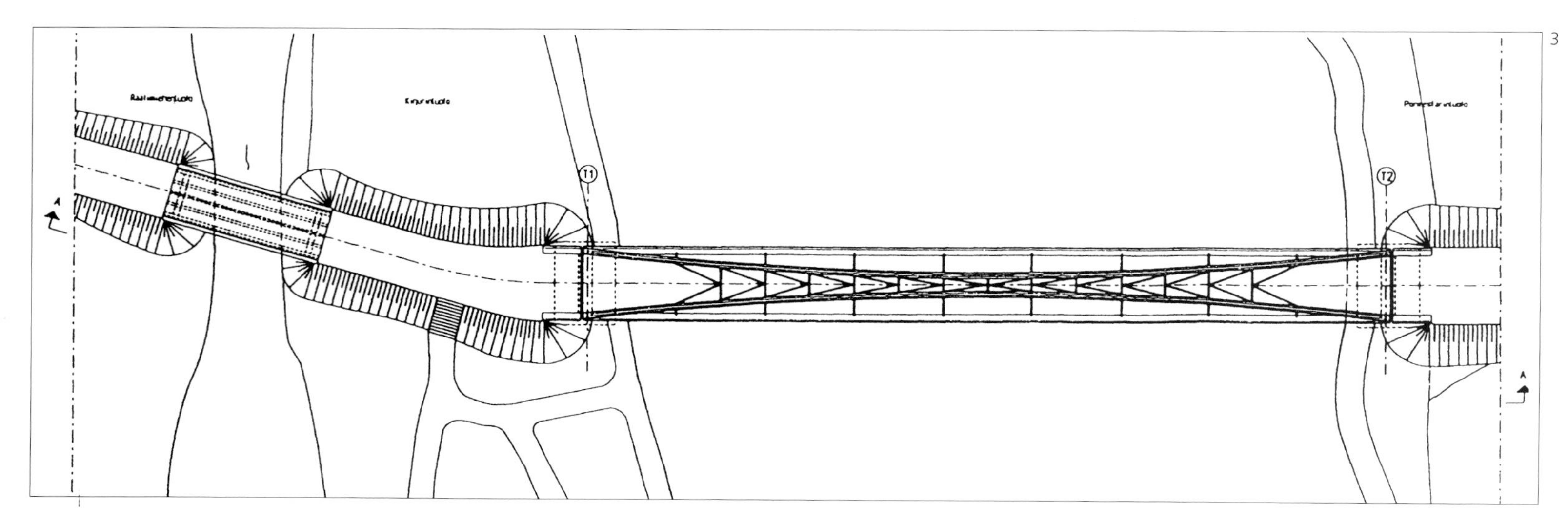

3

5 透视
6 从岛上看
7 河上的桥
8 该项目落成典礼

与尤哈尼·哈伏纳，彭泰克工程公司合作

5 Perspective
6 View from the island
7 Bridge over water
8 Inauguration day

In collaboration with Juhani Hyvönen, Pontek Engineers Ltd

5

6

7

8

赫尔辛基火车站
站台顶棚
2001

Helsinki Railway Station Platform Roofing 2001

赫尔辛基火车站建于1919年。火车站的建筑设计师沙里宁早在车站建设期间就曾经为站台设计了三种不同的顶棚方案，但没有一种被采纳。

现在的设计是根据1994～1995年建筑设计竞赛获胜方案发展而来的。在这个基础上根据甲方的意愿对原方案进行了更详细的修改。

站台顶棚的大小在很大程度上由原来火车站的规模所决定。端头16m×69m的顶棚覆盖了站台的主体区域，即南边与车站大厅毗连的地方。站台顶棚的北端建了一面与顶棚下表面齐高的墙。覆盖4—11号站台的顶棚较高，总长165m，这一部分站台为长途列车和中转列车的旅客服务。先进的高速列车Pendolino可以停靠在这里。

端头站台的顶棚建造在钢桩柱之上，墙边的柱子与原有建筑混凝土墙上的半柱结合在一起。墙和屋顶材料是玻璃。承重结构为钢。

站台顶棚的钢结构和钢拉杆都是喷漆钢。另外，尽端部分和路轨上方的顶棚采用了钢化玻璃。灯光系统和公告用的电子设备安装在建筑的实体部分。

Helsinki Railway Station was opened in 1919. While the station was under construction, its architect Eliel Saarinen designed three different schemes for roofing over the platform area, none of which were built.
Commuter traffic has increased dramatically over the last few years and the standard of service in the station generally had to be stepped up to match the new, rapid Inter-city express trains. In addition, weather conditions have always produced problems, especially in wintertime.
The design is based on winning entry in an international competition held in 1994-1995. The competition was open, but ten foreign architects were specifically invited to take part: Santiago Calatrava, Terry Farrell, Sir Norman Foster, Knud Holscher, Toyo Ito, Rafael Moneo, Jean Nouvel, Niels Torp, Volkwin Marg and Volker Giencke.
The dimensioning of the canopy over the platforms is tied in with Saarinen's existing station building. The canopy over the waiting area at the south end of the platforms is bounded by the main hall with its kiosks. The Pendolino trains, which represent the latest addition to Finland's rolling stock, fit in under the high platform canopy for their entire length.
The new waiting hall at the end of the platforms was built in the first phase as an unheated, roofed space, open to the platforms. The canopy over the waiting area is founded on driven steel piles, except for the columns adjacent to the walls, which are supported by the pilasters in the concrete walls of the station buildings. The walls and the roof of the canopy are of glass and the loadbearing structure is of steel. The south wall of the canopy is fitted with smoke vents. A frameless, point-fixed system was chosen for roofing the waiting hall, and a structural-glazing system fixed on a steel frame, was used for the high roof over the platforms. Annealed, clear, sheet glass is used in the waiting hall.
In the second phase, the area extending from the waiting area 165 metres to the north was roofed with a high platform canopy which completely covers the entire platform area.
Between the platforms, narrow slots that are themselves roofed over, have been left in the canopy, so that rainwater can drain down and soak away in the track area. The canopy is located above the overhead, electrified pick-up cables and between the first and second floor windows of the station buildings. The steel structure of the canopy including the tension rods is painted.
The first section of the roof over the platform area and the glass at the edge of the east wing are glazed with electrically-heated, annealed, laminated glass. The non-glazed parts of the canopy house the electrical installations for the lighting and public address systems.
The canopy is fitted with steel-mesh platforms and trolleys for maintenance purposes, and the edges of the canopies are fitted with opening lights to provide access for removing excess snow. Ventilation and smoke extract are both handled by natural circulation.

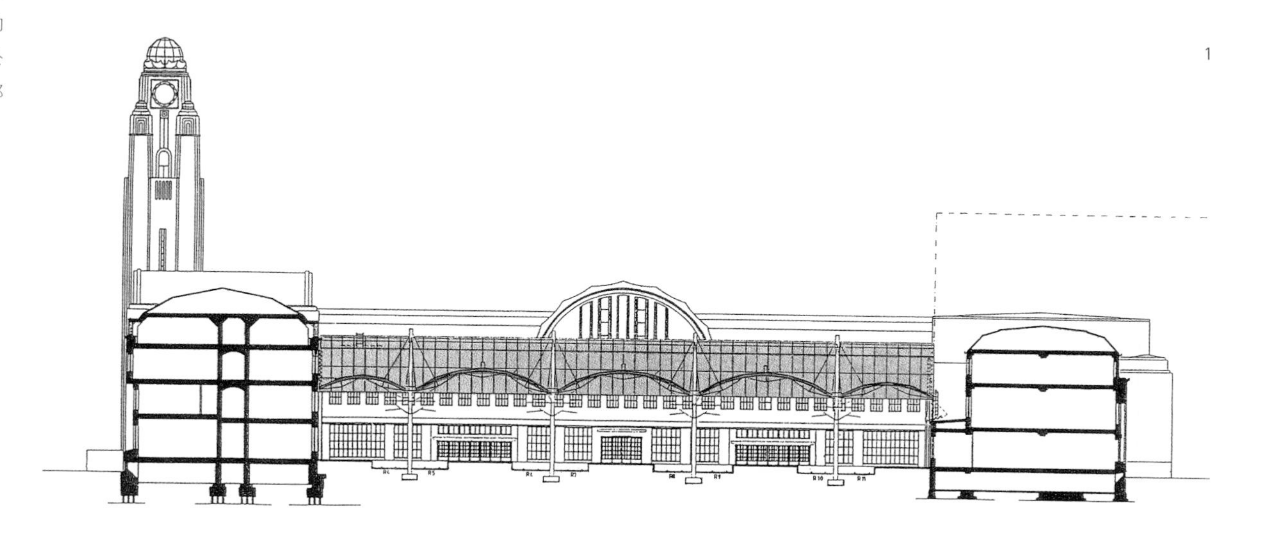
1

1 剖面

1 Section

2 在赫尔辛基市中心的位置
3 屋顶平面
4 剖面
5 细部

2 Situation in the centre of Helsinki
3 Roof plan
4 Section
5 Detail

2

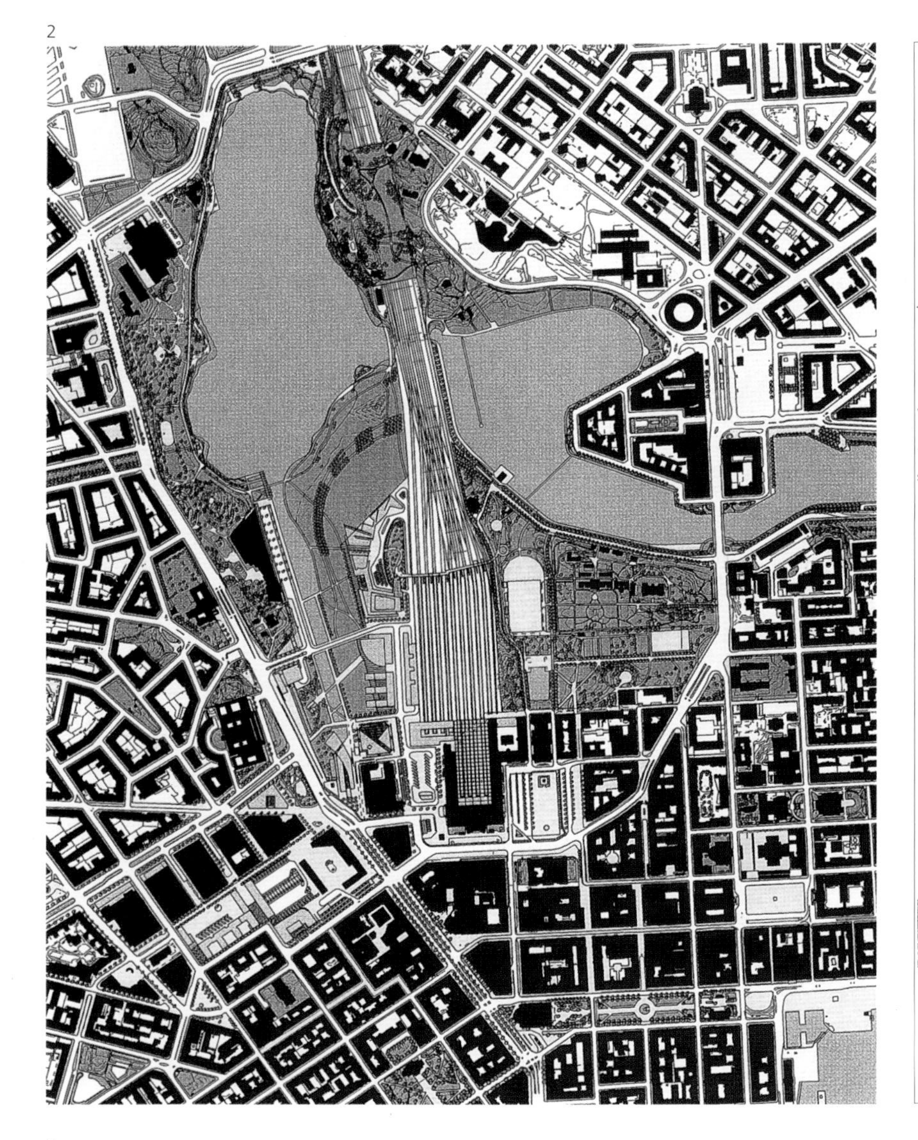

3

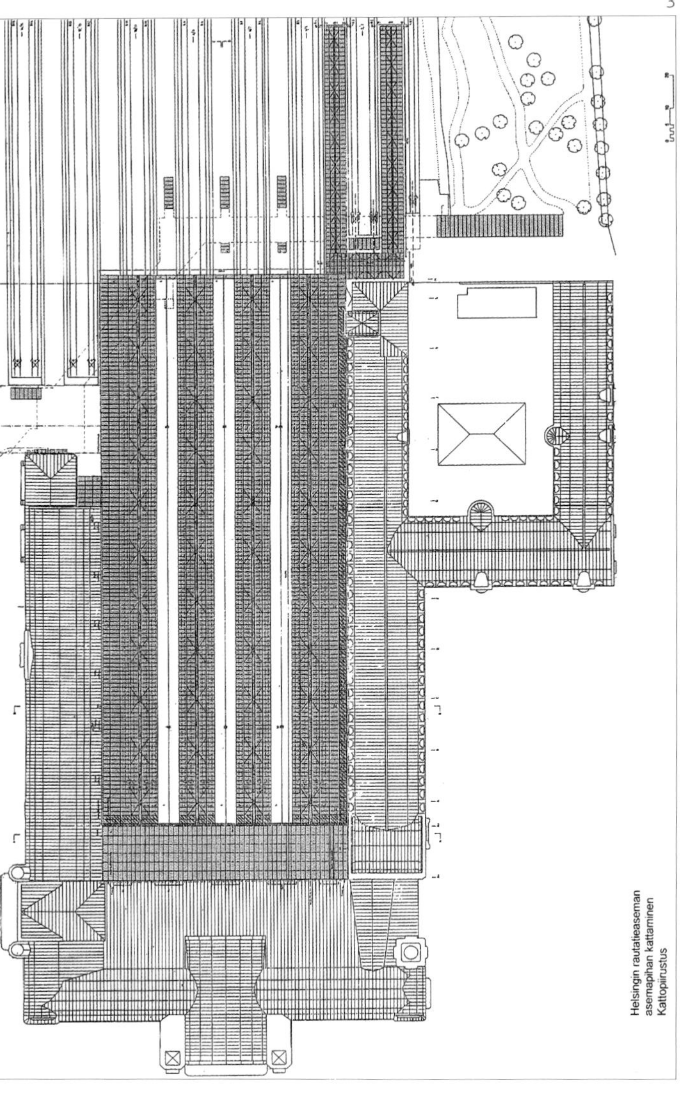

4

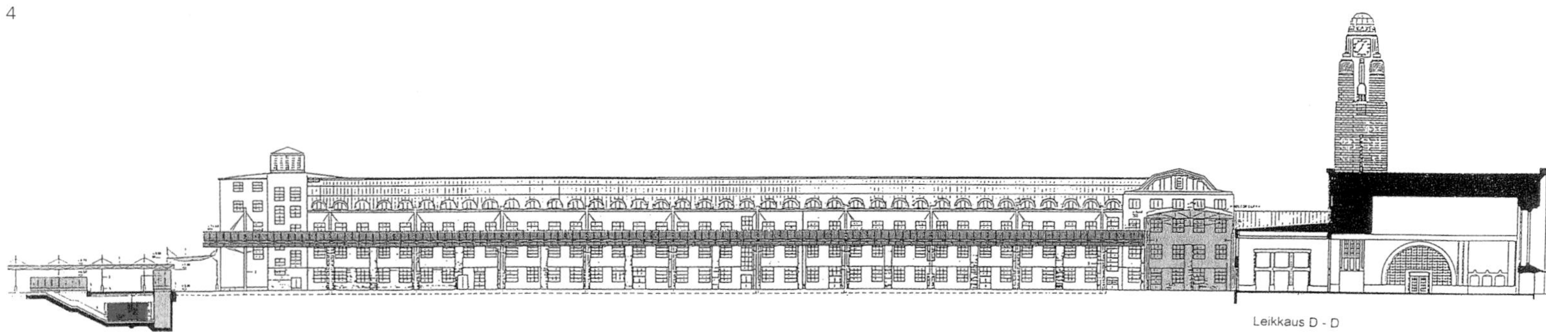

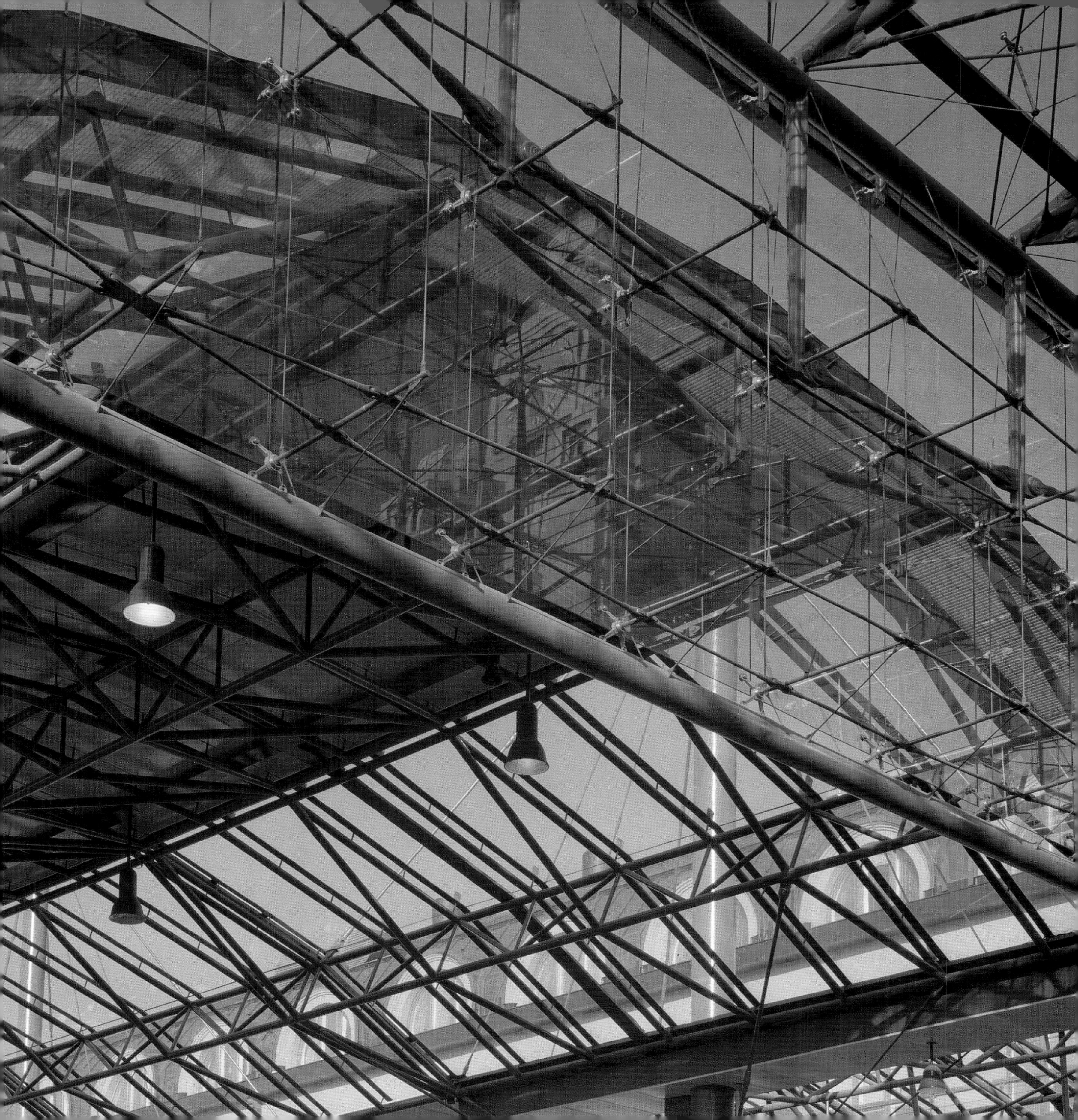

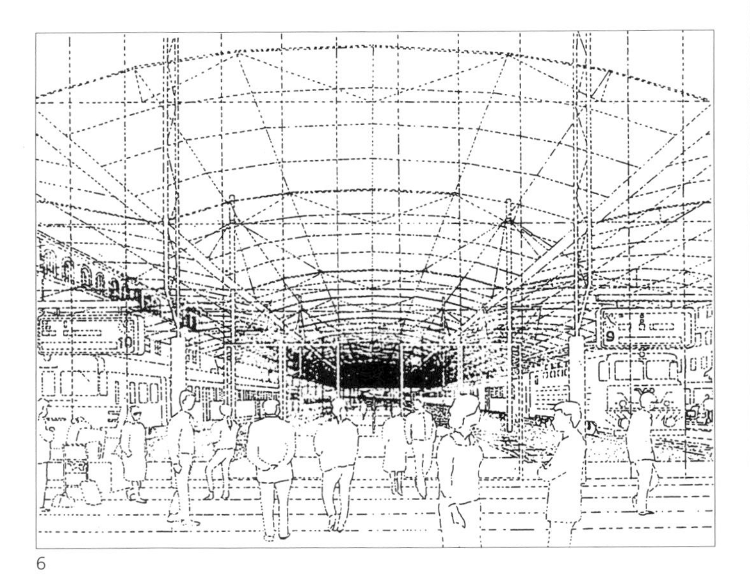

6

7

6 透视
7 候车月台端部

6 Perspective
7 End of the platform area

Helsinki
Helsingfors
aina tuoretta nopea

8 模型照片
9 剖面
10 细部
11 候车平台终端
12 候车平台终端
13 候车平台高雨棚

8 Model photo
9 Section
10 Details
11 End of the platform area
12 End of the platform area
13 High platform canopy

8

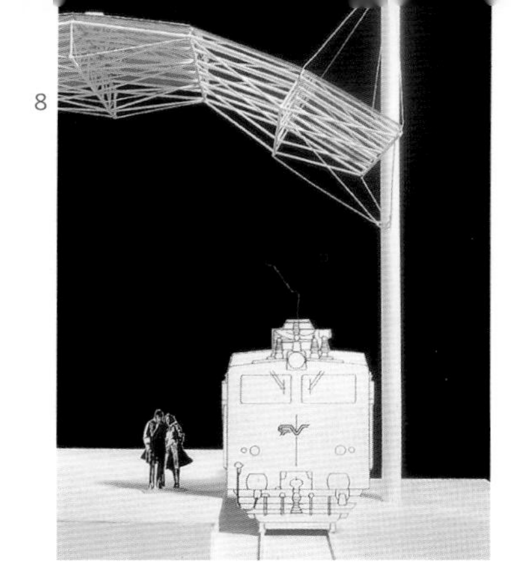

9

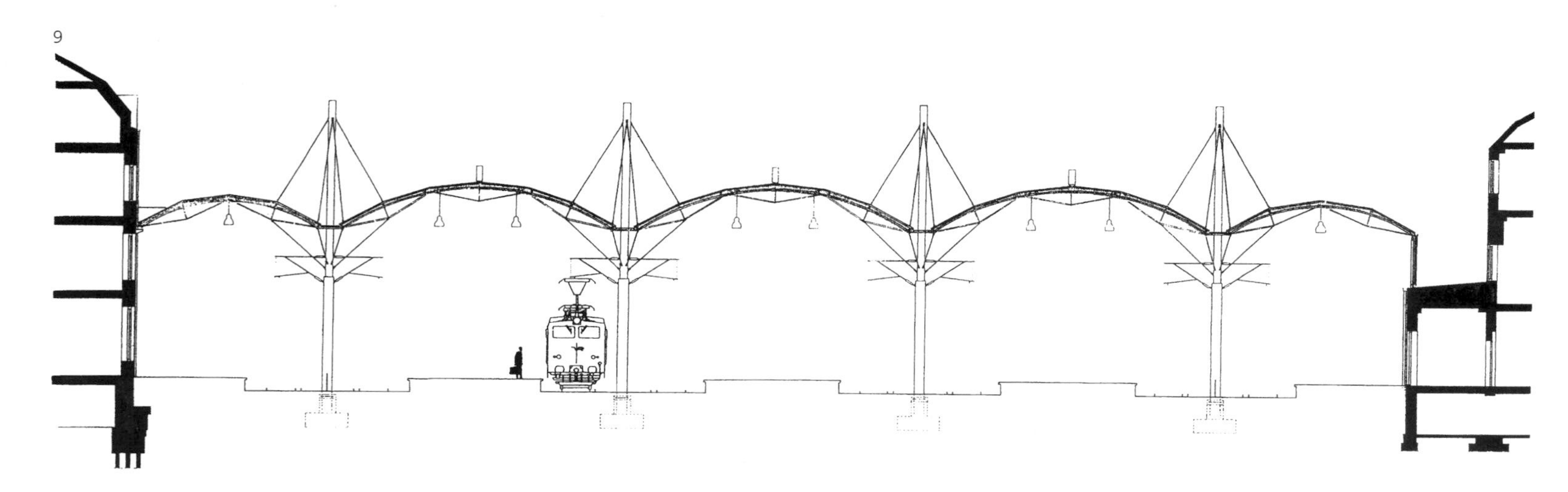

10

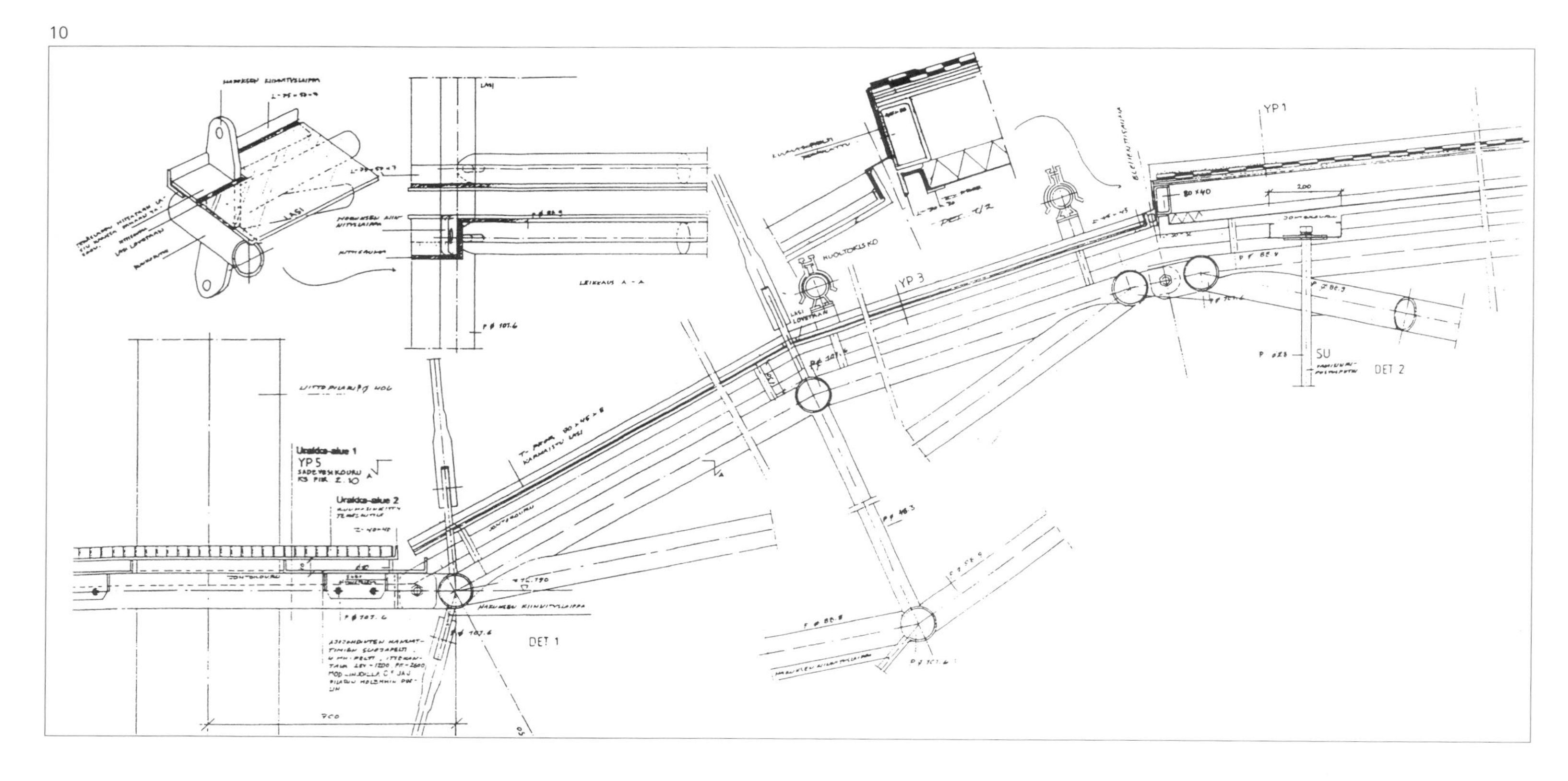

11

12

15:58
Oulu
Tampere Seinäjoki
7
16:02
Tampere Tammerfors
6
26118
2364

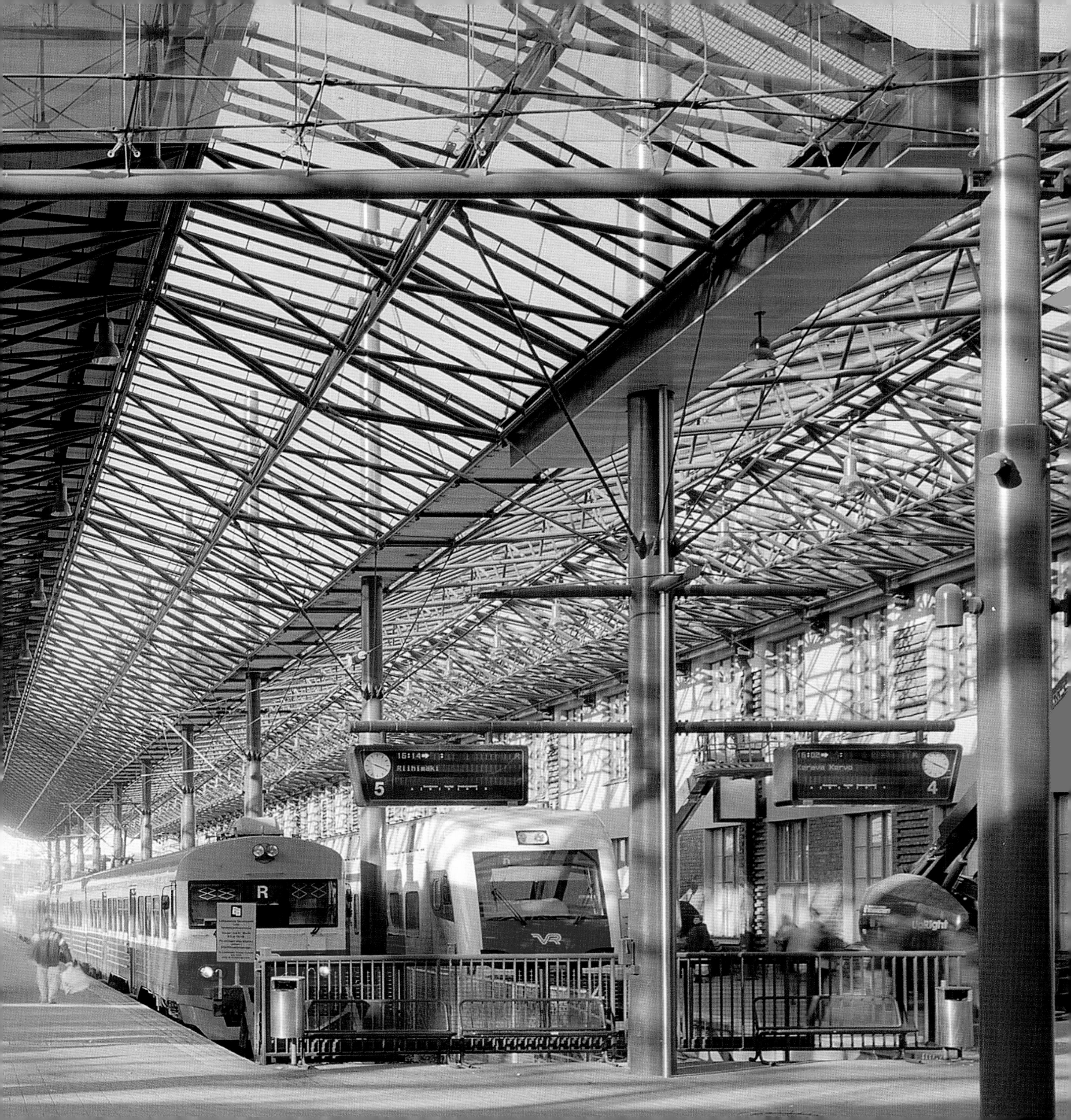

16:14
Riihimäki
5
16:02
Kerava Kervo
4
R
VR
UpRight

14 钢格
15 细部
16 细部
17 透过玻璃顶看钟楼
18 来自钟楼的视点
19 候车平台高高的雨棚

14 Steel grid
15 Detail
16 Detail
17 Bell tower seen through the glass roof
18 View from the bell tower
19 High platform canopy

14

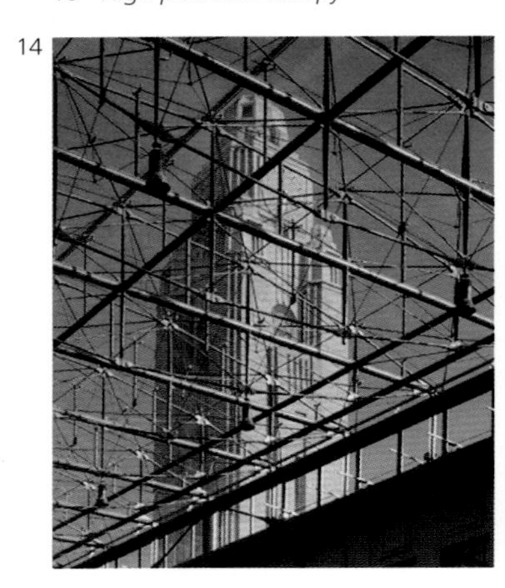

17

15

18

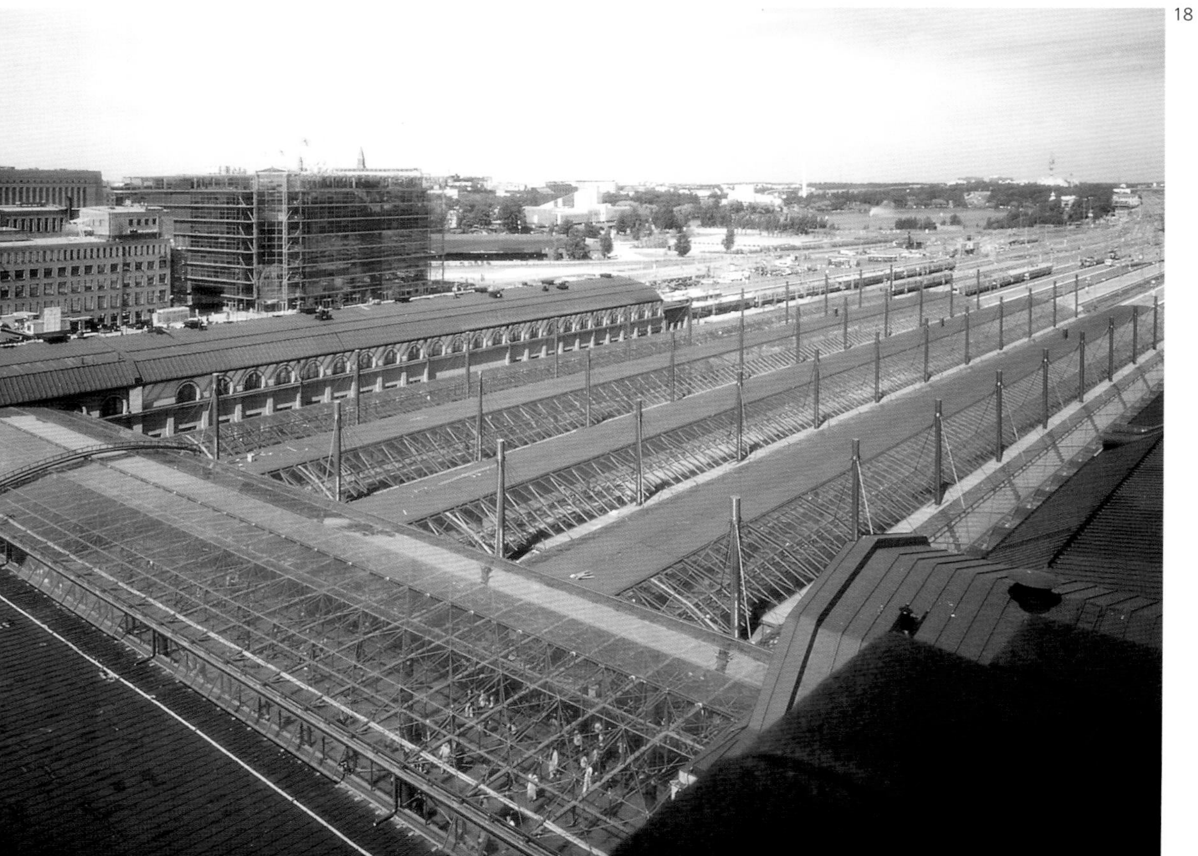

16

20 钢结构网格
21 细部草图
22 细部
23 细部
24 细部
25 细部
26 剖面
27 端部侯车月台雨棚

20 Steel construction grid
21 Detail sketches
22 Detail
23 Detail
24 Detail
25 Detail
26 Section
27 End platform canopy

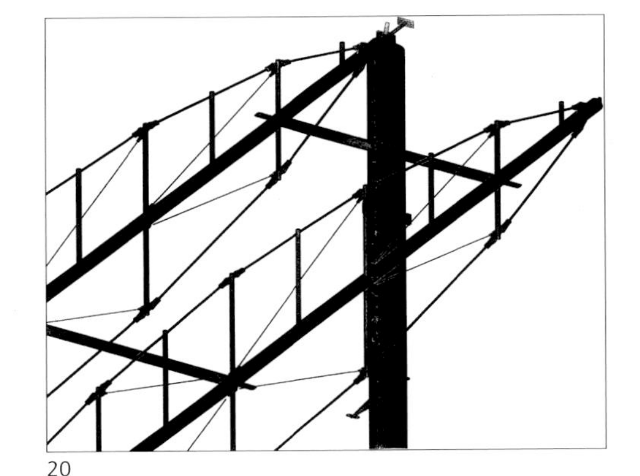

20

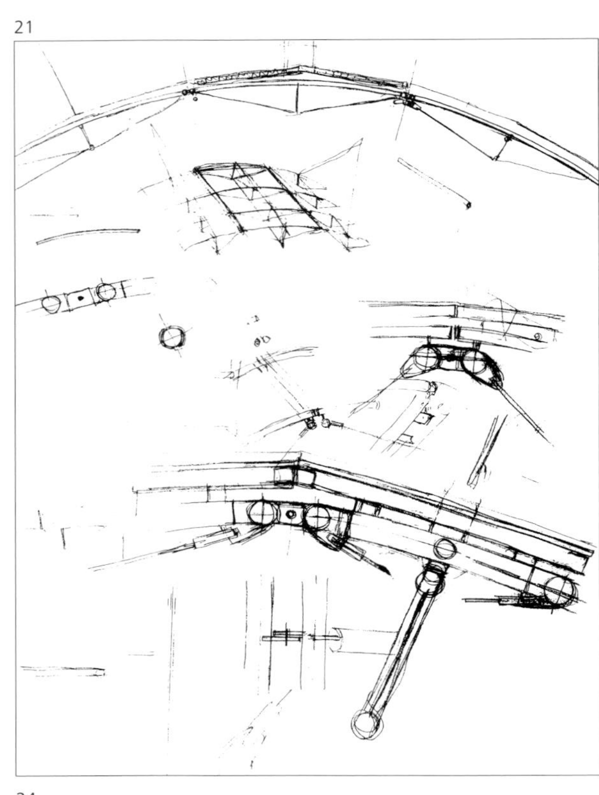

21

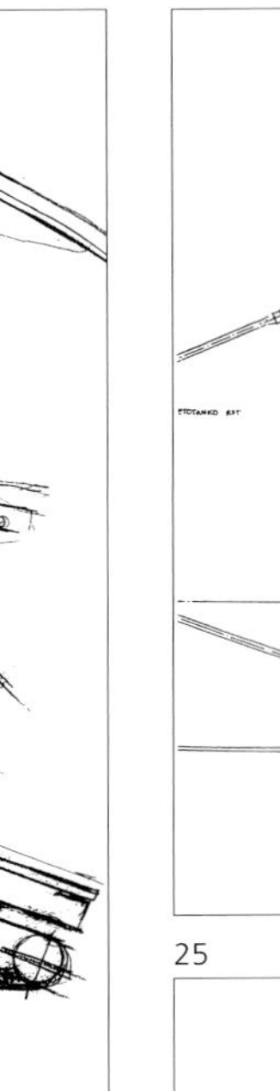

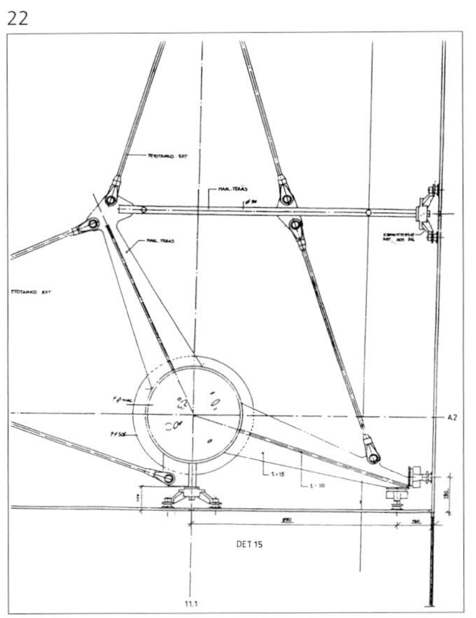

22

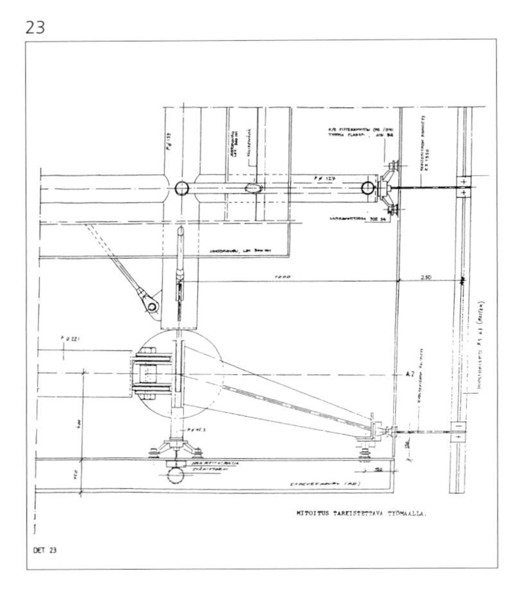

23

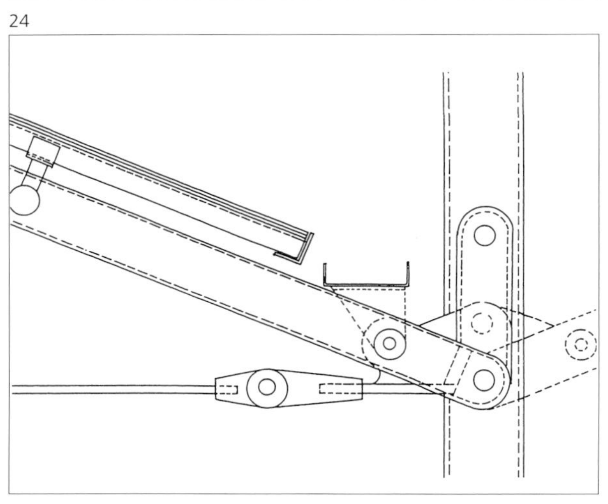

24

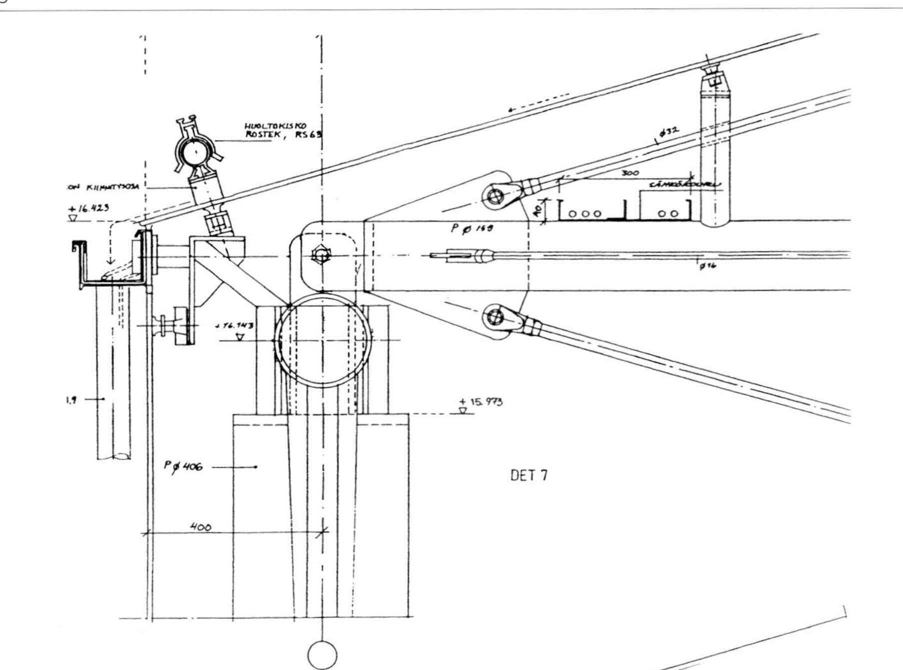

25

26

Helsingin rautatieaseman
asemapihan kattaminen
Leikkaus A - A
B
A
A_2
A_1
0
5

27

28 轴测图
29 连接
30 模型照片
31 从北面看候车平台顶
32 从候车月台顶部上方看

28 Axonometric
29 Joint
30 Model photo
31 Platform roofing from the north
32 View above the platform roofing

28

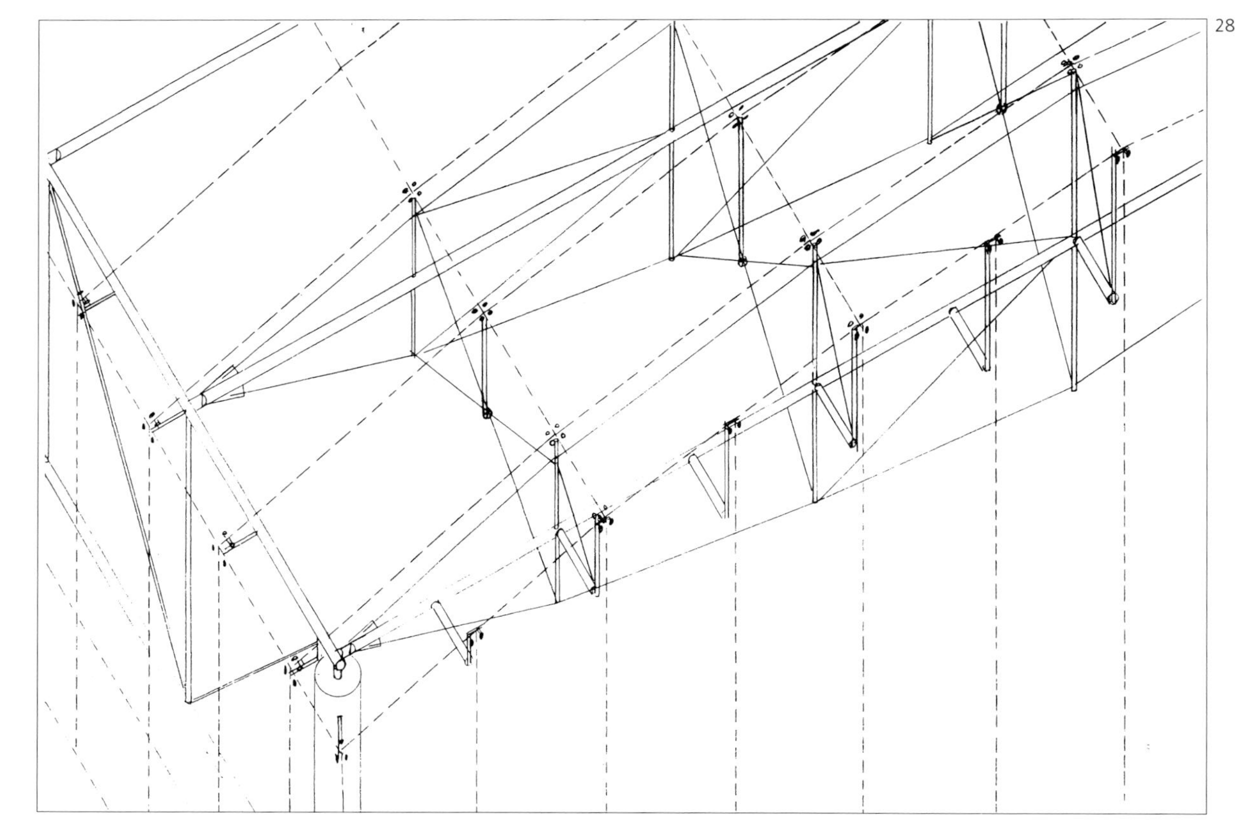

29

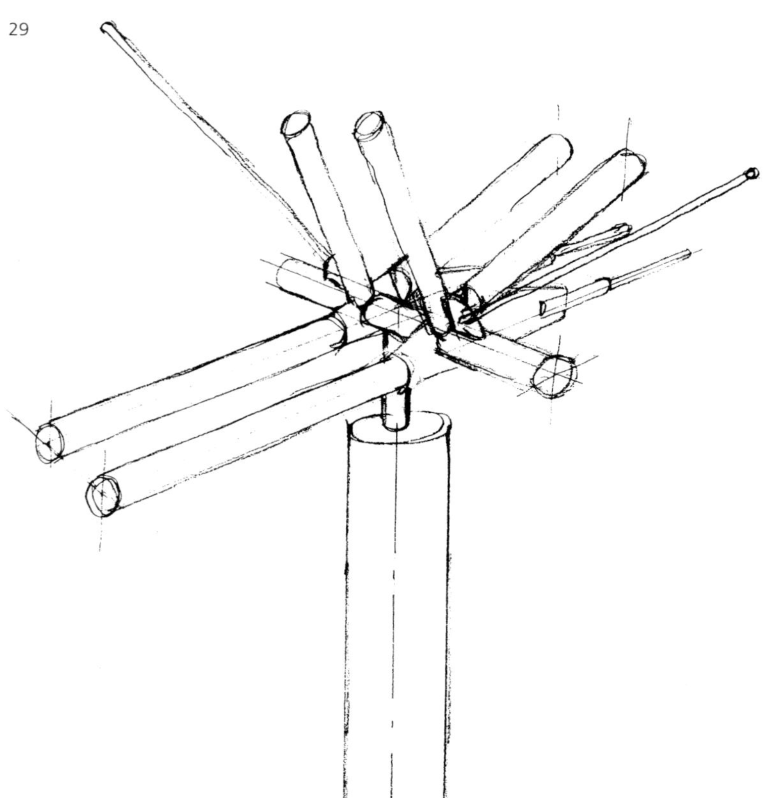

30

31

展览设计　Exhibition design

Taidetapahtuma 1 展览，图尔库 1966

Taidetapahtuma 1 Exhibition, Turku 1966

1966 年图尔库举行了第一次学生艺术节。不同领域的艺术通过诸如音乐会、讨论会、展览会、戏剧表演等方式展示出来。

视觉艺术的展览被安排在图尔库艺术博物馆。

我们被指定设计包括海报、宣传册和作品集等在内的所有印刷品。

Students Art Festival in Turku was arranged for the first time in 1966. All fields of art were represented in shows, concerts, seminars, exhibitions, theatre performances etc.
The exhibition of visual arts was arranged in Turku Art Museum.
We were asked to design also all the printed materials like posters, brochures and anthology.

1 展台
2 展台

1 Exhibition stands
2 Exhibition stands

1

2

艾里克 · 布里格曼展览，图尔库 1967 赫尔辛基 1968

Erik Bryggman Exhibition, Turku1967 Helsinki 1968

1 邀请卡封面
2 展板与模型
3 展板

1 Invitation card cover
2 Exhibition panels with model
3 Exhibition panels

众所周知，二十世纪二十年代末，建筑师艾里克·布里格曼和阿尔瓦·阿尔托一样，是芬兰倡导实用主义的社会和形式理念的先锋。作为一个先行者，布里格曼本人并没有遵从实用主义中的教条原则，他的作品里总是非常清楚地体现出一种柔和的、非常诗意化的追求，这一点可以从他那著名的作品——图尔库复活教堂的室内设计中表现出来。

艾里克·布里格曼回顾展由图尔库市主办。

It is generally aknowledged that Erik Bryggman, the architect, together with Alvar Aalto, were the pioneers in promoting the social and formal ideas of functionalism in Finland in the late 1920s. Although being the predecessor, Bryggman did not obey the strictest principles of functionalism, but in his works there is a main stream of a softer and more poetic hold, a result of which is the famous interior of The Resurrection Chapel in Turku.
The retrospective exhibition of Erik Bryggman was sponsered byt the City of Turku.

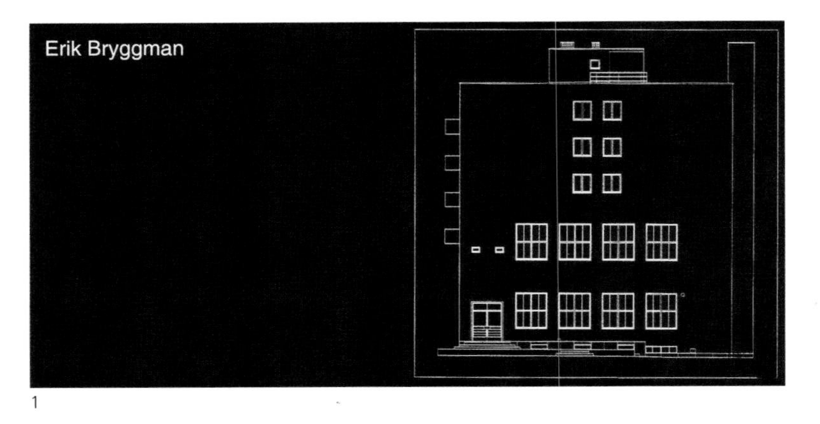

1

2

3

街道上的设施展览，赫尔辛基 1970

Street Furniture Exhibition, Helsinki 1970

城市就像是一个舞台，需要有一些设施为来往的人群提供方便，为他们的行动提供引导。城市也要为一些偶然的但又是必不可少的活动提供庇护，这就需要房子、标识和广告、还有休息的地方等等一系列设备。这样的要求决定了街道和广场的公共性质——街道也因为这些活动而存在。

宣传用的信息板有些很小，而有些越做越大，直到最后失去了它们存在的初衷。街道由于广告牌的相互竞争，造成了视觉上的混乱。如果把休息亭、灯柱、照明设备、红绿灯和交通标识加进来的话，街道就成了一个规划不力的令人不愉快的地方，像一个垃圾场。

城市的公众空间必须被看成是城市整体的一个组成部分，它的规划必须要经过慎重考虑。

这是1970年在芬兰建筑博物馆举行的街道设备展。在这之后，在全芬兰巡回展出。

The city is like a stage which needs something to accommodate moving people, something to guide their activities. It needs shelters for occasional but indispensable actitivities, it requires constructions, signs and advertisements, places to sit on, a whole army of equipement. With these objectives the final character of streets and squares is determined - the street is born of their existence.
The signboards range from small information boards to bigger and bigger boards, until they finally destroy their original purpose. The street has become, because of billboards competing with one another, a visual chaos.
If kiosks, lamp posts, lighting devices, traffic lights and signs are added to this, the street may become a badly planned and unpleasant place, a heap of rubbish.
The public space of cities must be seen as an integral part of the city and its planning must be considered important.
Street Furniture Exhibition was held in the Finnish Museum of Architecture in 1970. After that it was shown around Finland.

1

1 展示想像图
2 展板

1 Image of the exhibition
2 Exhibition panels

2

G4 imago
展览
1977

G4 imago
Exhibition
1977

1

2

G4小组（由建筑师Ola Laiho Esko Miettinen，Juhani Pallasmaa 和Esa Piironen组成）的成员自从60年代早期开始，就已经把平面设计作为他们建筑设计以外的一份工作。G4 曾参加过下面的平面设计展：1966 年图尔库，1967年Espoo（Dipoli），1967年斯德哥尔摩和1977年赫尔辛基。

在赫尔辛基举行的展览的目的是向公众传达平面设计的重要性和当今工程的复杂性。

在G4的展出中，特别强调了设计者和民众共同的教育背景之间的相互关系。展览覆盖了平面设计的如下几个领域：企业形象，标识系统，展览设计，街道设备和产品设计。

G4 的作品已经发表在《Graphis Annual》、《Modern Publicity》、《Graphis Diagrams》和《Graphis Posters》等多种刊物上。该小组设计的海报于1969年在威尼斯双年展的艺术宣言展区展出。

The members of the group G4 (architects Ola Laiho Esko Miettinen, Juhani Pallasmaa and Esa Piironen) have, since the early 60s, done graphic design in addition to their architectural practice. G4 has had the following exhibitions of graphic design: Turku 1966, Espoo 1967 (Dipoli) and Stockholm 1967 and Helsinki 1977.
The aim of the exhibition held in Helsinki was to inform the public about the importance of graphic design and the complexity of the projects today. By the exhibition G4 emphasized the importance of the co-operation of designers and the common background of education as well. The exhibition covered the following fields of graphic design: corporate images, signing systems, exhibition design, street furniture and product design.
The works of G4 have been published in several issues of Graphis Annual, Modern Publicity, Graphis Diagrams, Graphis Posters. Posters designed by the group were exhibited in Manifesto d'Arte in the Venice Biennale in 1969.

3

4

1 帕拉斯玛设计的展览标志
2 展台
3 展台及展板
4 展台及展板

1 Symbol for the exhibition by Juhani Pallasmaa
2 Exhibition stands
3 Exhibition stands and panels
4 Exhibition stands and panels

Lars Sonck 展览 1981

Lars Sonck Exhibition 1981

1

2

Lars Sonck（1870—1956）是芬兰20世纪建筑界的关键人物之一，他的名字也被全世界所知晓。这个展览介绍了Sonck作品的整个发展过程，从国家浪漫主义先锋到较为纯净的纪念主义风格。展览的小册子中包括了Lars Sonck较为详细的生平传记和作品目录。展览不仅在芬兰的各个城市举行，同时也在伦敦、西雅图、华盛顿、明尼阿波利斯、纽约和芝加哥等地举行。

Lars Sonck (1870-1956) was one of the key figures of 20th century Finnish architecture whose name is also known abroad. The exhibiton presents the main lines of the development of Sonck' works from the breakthrough of National Romaticism towards a more simplified monumentalism. The exhibiton brochure includes a more detailed biography and a catalogue of his works. The exhibition was shown in most Finnish cities as well as in London Seattle, Washington DC, Minneapolis, New York and Chicago.

1 展台
2 展台
3–6 平面
7 展台

1 Exhibtion stands
2 Exhibition stands
3-6 Floor plans
7 Exhibition stands

3 4

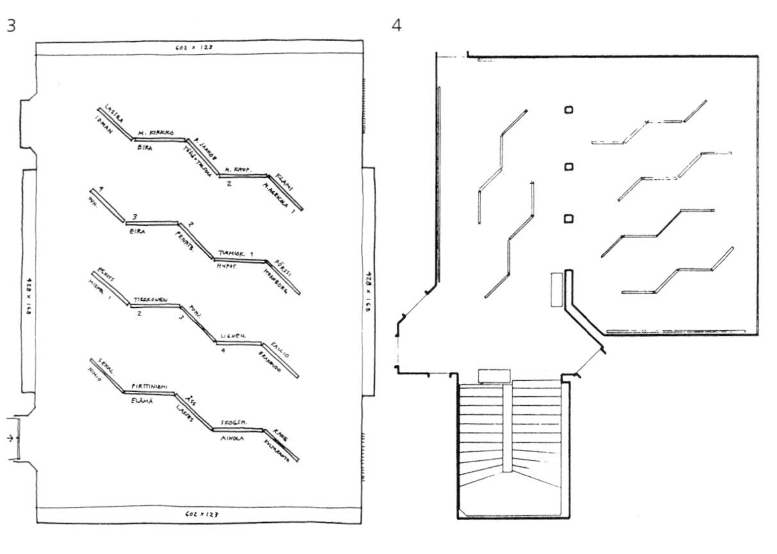

5

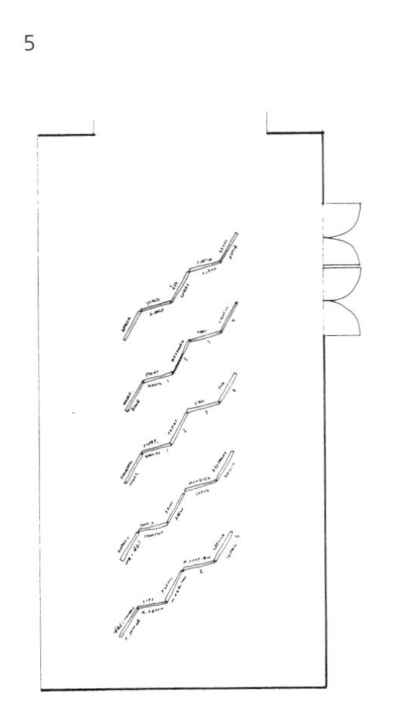

6

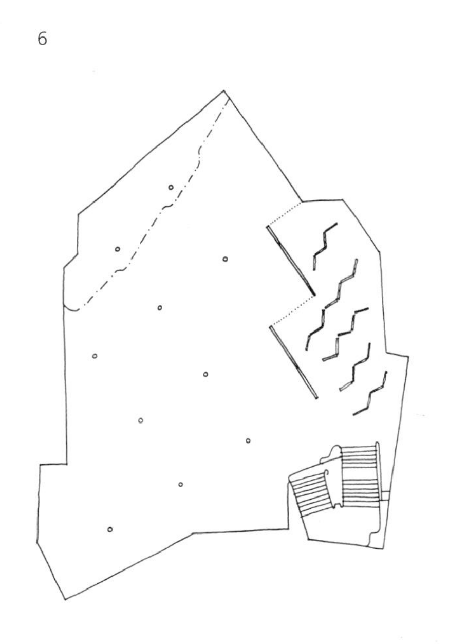

7

Ars Sacra Fennica 展览，赫尔辛基 1987

Ars Sacra Fennica Exhibition Helsinki 1987

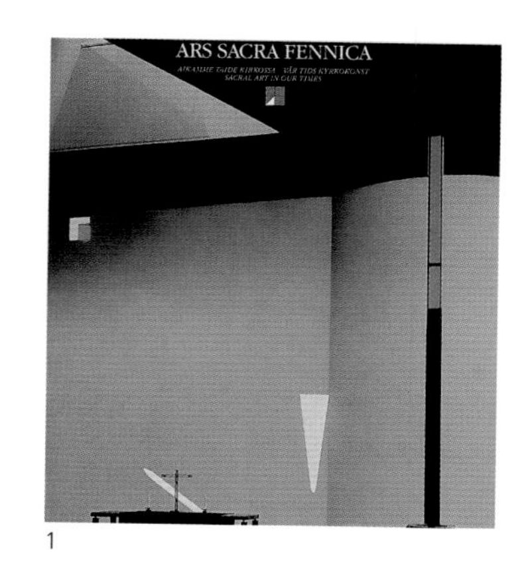

1

2

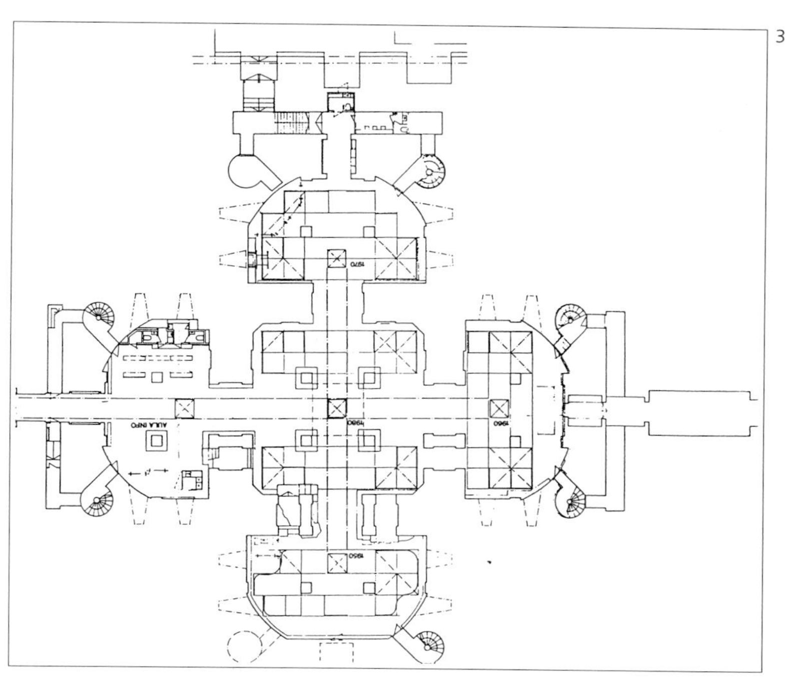

3

1987 年，在赫尔辛基大教堂的地下室举行了一个以战后宗教艺术为主题的大型展览，类似的展览过去从未举行过。

展览的内容主要是过去几十年间在宗教建筑、绘画、教堂设备和纺织品等领域创作的作品。

二战后，芬兰大约建造了200座教堂，其中的许多是由阿尔托、塞任佩特拉等享有国际盛名的建筑师设计的。教堂设计是一个很大的挑战，然而在这个领域里尝试过的许多年轻建筑师同样也获得了成功。展览表现了现代建筑的许多方面。

新的教堂建筑已经能够被人们接受和承认，但它们还并不能被承认为艺术品。在很长的一段时间里，芬兰的教堂非常缺乏视觉艺术的设计。战后新建的教堂中，半数的室内仅仅只用十字架作为装饰，其余的则摆放了一些传统风格的耶稣受难像、油画、雕刻和纺织品等。抽象的作品几乎没有。

50 年代末见证了一种反禁欲主义的声势浩大的浪潮，现代主义唤醒了情感的冲突和广泛的公众讨论。

一些教堂尝试用新的方式进行装饰，但任何偏离传统做法的行为常常遭到冷漠、甚至敌意的对待。

宗教艺术在近年来分支出很多新的领域。大量的竞赛吸引了一些非常优秀的设计师，为我们的教堂设计出更好的艺术作品。

与 Jyrki Nieminen 合作
1 展览标志
2 展览景观
3 平面

A grand exhibition of sacral art in post-war Finland such as never shown before was held in the crypt of Helsinki Cathedral in 1987.
It was devoted to the major works of art produced in the last few decades in the fields of architecture, painting, church equipment and textiles.
Almost two hundred churches have been built in Finland since the Second World War, many of which designed by architects of international repute, such as Aalto, Sirén and Pietilä. Designing a church presents a great challenge and it is within this field that many of our younger architects have earned fame as well. This exhibition illustrates the many facets of modern architecture.
New church architecture has been received and accepted with relative ease, which cannot be said of the works of art. For a long time there was a conspicuous absence of visual art in Finnish churches. About half of the churches built since the war are adorned only by a cross. The rest have crucifixes, paintings, sculpture and textiles which are in the traditional style for the most part. There are few abstract works.
The late 1950s witnessed a reaction to this asceticism in a bold wave of modernism that aroused conflicting emotions and widespread public dispute.
Some churches were decorated in a new way, but any divergence from the familiar practice often met with indifference and even hostility.
Sacral art in recent years branched out into new fields. A number of competitions attracting some of the very best artists have been held to achieve works of art and textiles for our churches.

in collaboration with Jyrki Nieminen
1 Symbol for the exhibition
2 View of the exhibition
3 Floor plan

设计　Design

不锈钢烛台
1978

Steel Candlelight Fixtures
1978

这个系列设计的主题是"永恒"，它由不同用途的蜡烛架构成。蜡烛固定器的尺寸不同。材料是不锈钢的。这个系列目前只设计了几种样板，没有投入大规模生产。

The series, which represents timeless design, consists of candlesticks for different purposes. There are holders for candles of different sizes. The material used was stainless steel. Only a few items of the series were made and it never reached a large market.

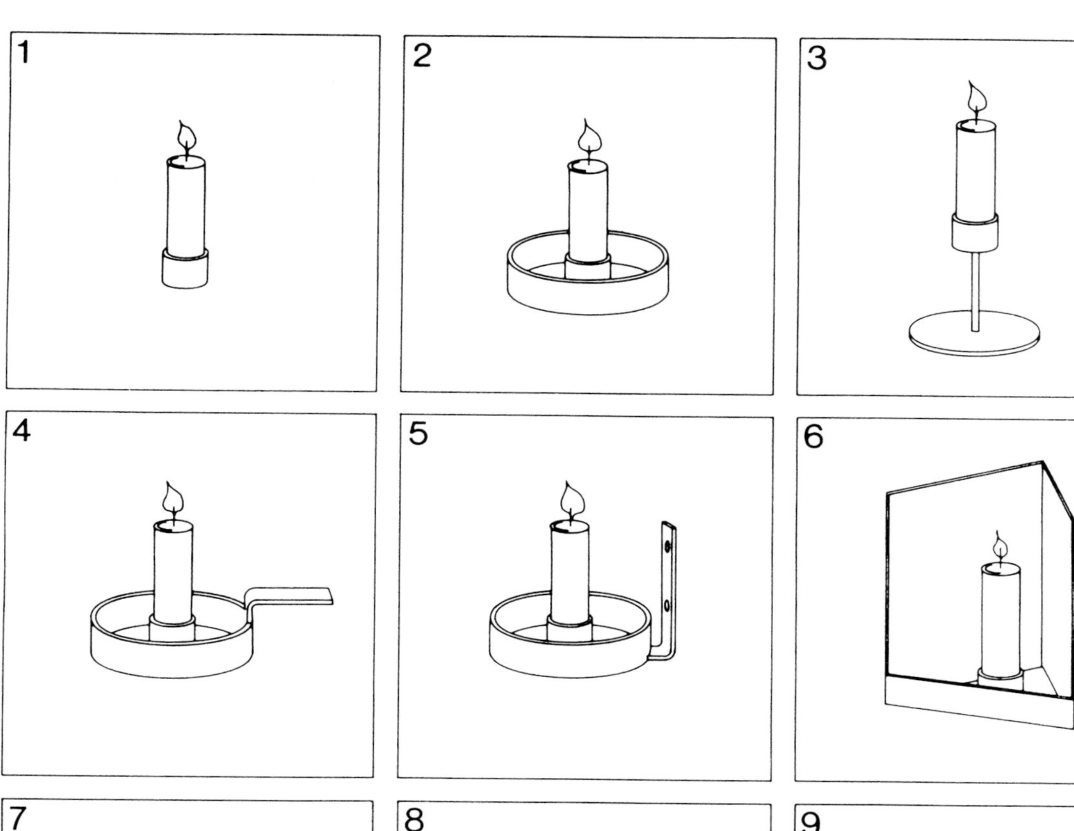

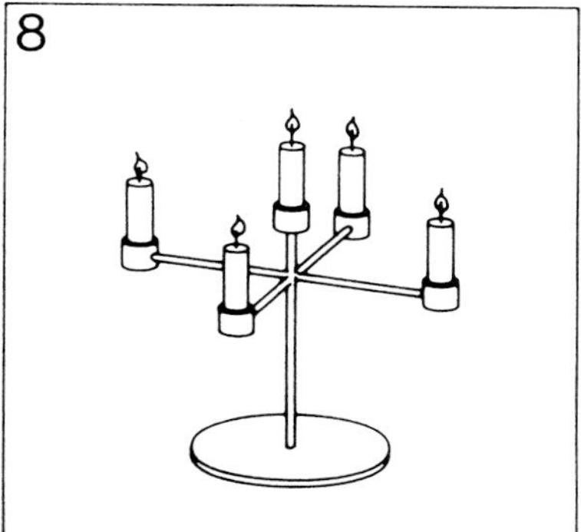

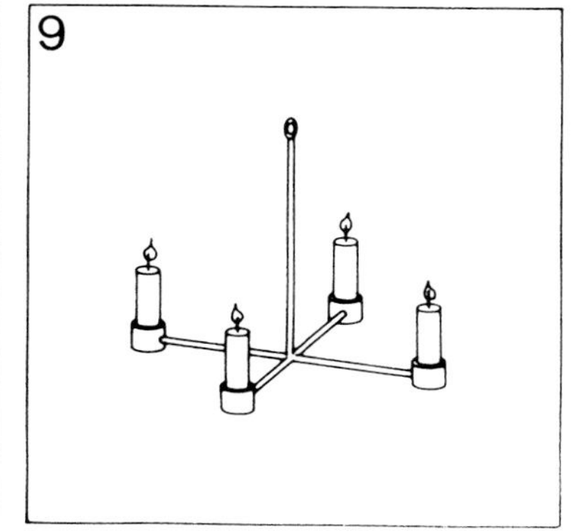

1 试制模型
2 各种不同模型

1 Prototype
2 Various models

公共汽车站台
Espoo
1992

Bus Shelter
Espoo
1992

1 模型照片
2 竞赛制图
3 竞赛制图
4 细部
5 试制模型

1 Model photo
2 Competition entry drawings
3 Competition entry drawings
4 Details
5 Prototype

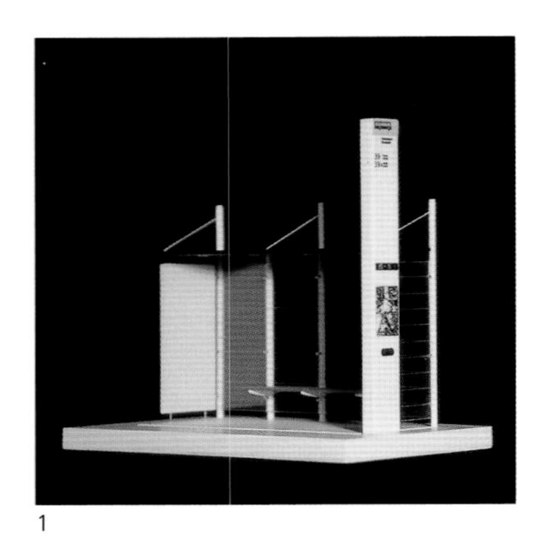

1

公共汽车站台设计的基础概念是部件组合，即通过局部部件的替换获得多种多样的变化结果。设计采用了高质材料和精巧细节，目的是为了抵制故意破坏的行为。

站台包括一个信息塔、供候车用的长凳、垃圾箱、照明设备、供热单元以及用来显示公交车运行车次的电子显示屏。大部份信息面板都有背景照明，这样晚上也可以看得清楚。

公共汽车站台内也可以设电话亭和广告。但站台外观总地看起来是通透的。

站台使用的材料是不锈钢、8mm厚钢化玻璃和强化塑料。

公共汽车站台依据的原型是1992年由Espoo市、芬兰地区管理协会和芬兰住宅博览会举办的设计竞赛的获奖作品。

The design of the bus shelter is based on a concept of components, which makes it possible to create variations of the different parts. High-class materials and refined details are used as a means of discouraging vandalism.
The technical components of the shelter include an information pylon, bench, rubbish bin, lighting appliances, infra-heating elements as well as electronic monitors showing the running times of public transport vehicles. Most of the information panels are provided with background lighting making them visible at night as well.
The bus shelter may also include a telephone booth and advertisements. Transparency, however, remains an essential element of the shelter's appearance.
The materials of the shelter are stainless steel, 8mm toughened glass and reinforced plastic.
The prototype of the bus shelter was built on the basis of the winning entry in the design competition arranged in 1992 byt the City of Espoo, the Association of Finnish Local and Regional Authorities and the Finnish Housing Fair.

2

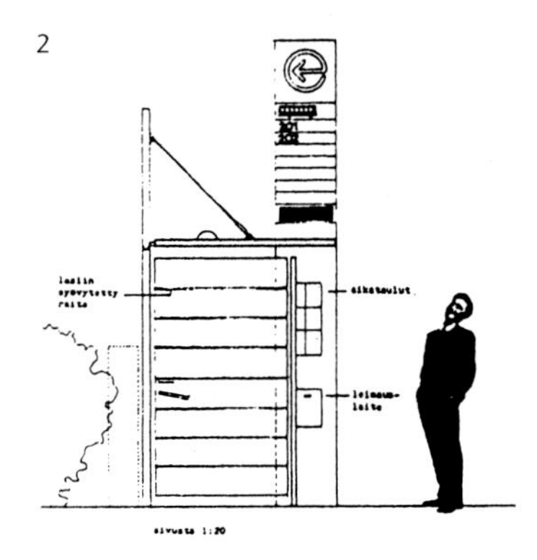

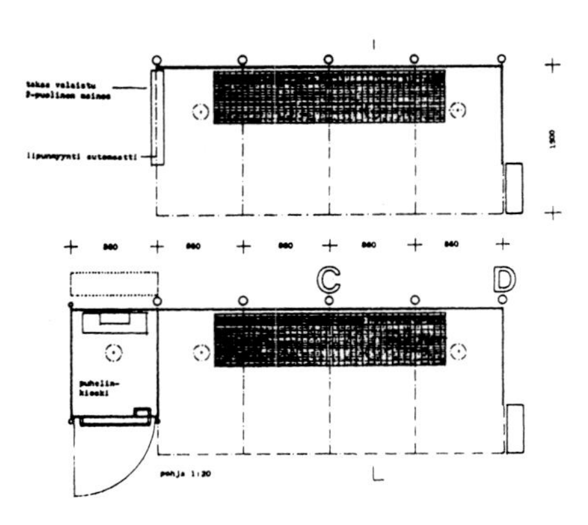

3

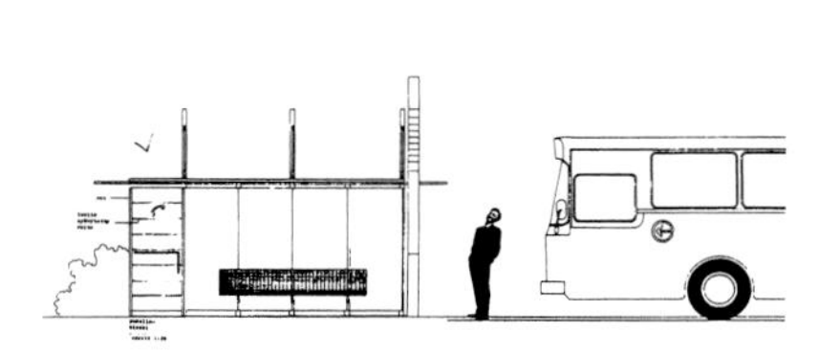

4

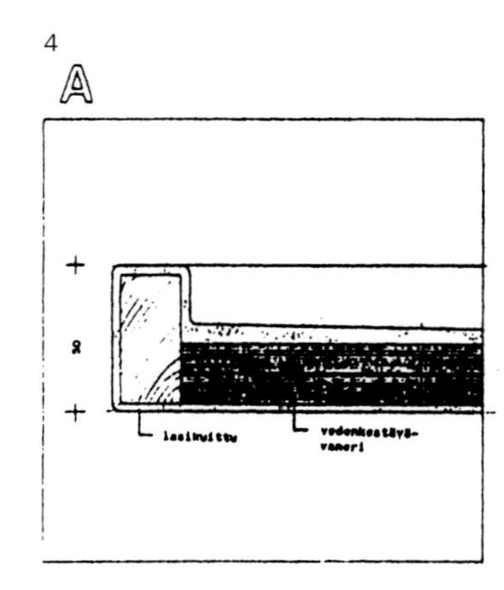

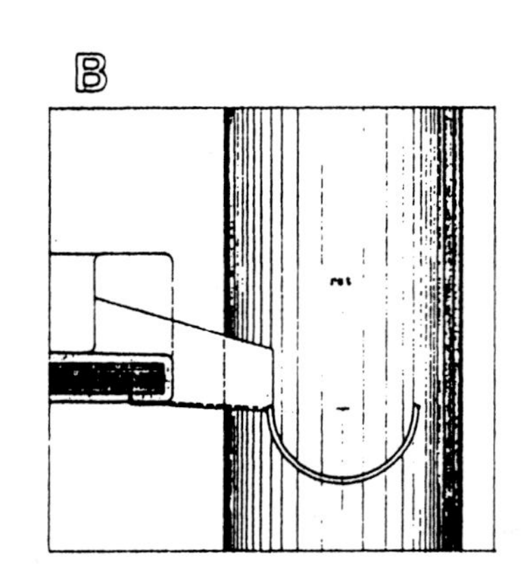

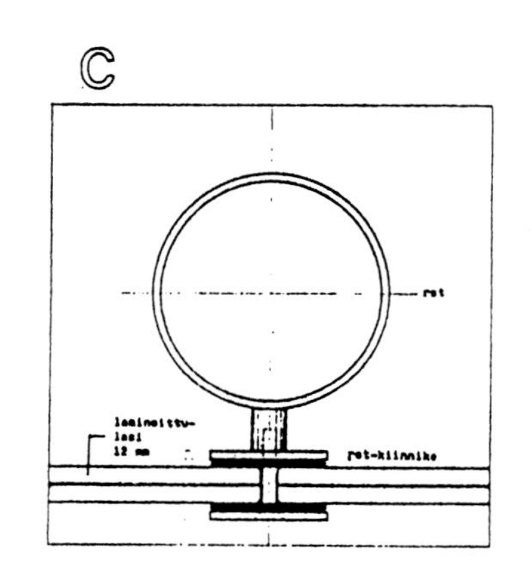

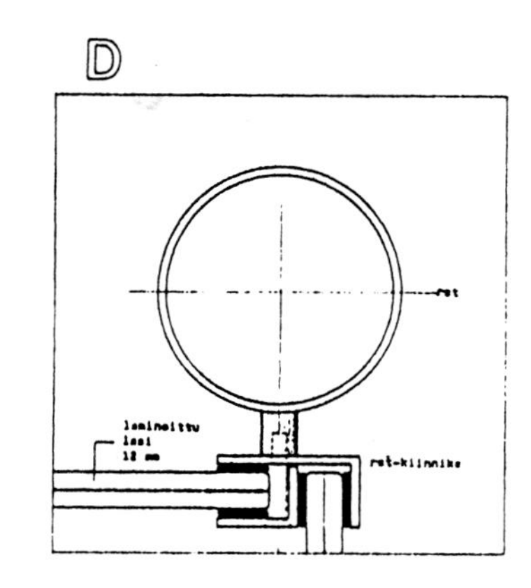

Tapiontori
Tapiotorget
10
14
16
Matkalippujen hinnat
pääkaupunkiseudulla
1992
14
5

Hietaniemi 公墓灯光设备，赫尔辛基 1993

Hietaniemi Cemetery Lighting Fixtures Helsinki 1993

Hietaniemi 公墓主通道的照明工程与其他的一些技术更新工程一同进行。

由于在市场上很难找到合适的照明设备，所以我们为公墓特别设计了灯柱。

灯光设备照亮了整个纵向通道。这里使用了合金灯。

与小教堂相反的方向，灯具位置较低，与公墓的气氛相符。

The lighting of the main aisle of Hietaniemi Cemetery was renovated together in connection with some technical construction work.
The lamp posts were designed specifically for this purpose, as suitable light fixtures could not be found on the market.
The ligth fixture lights up the aisle longitudinally. Multimetal lamps have been used here.
Opposite the chapel especially for this purpose designed low fixtures were installed, which are also supposed to give the cemetery a proper feeling.

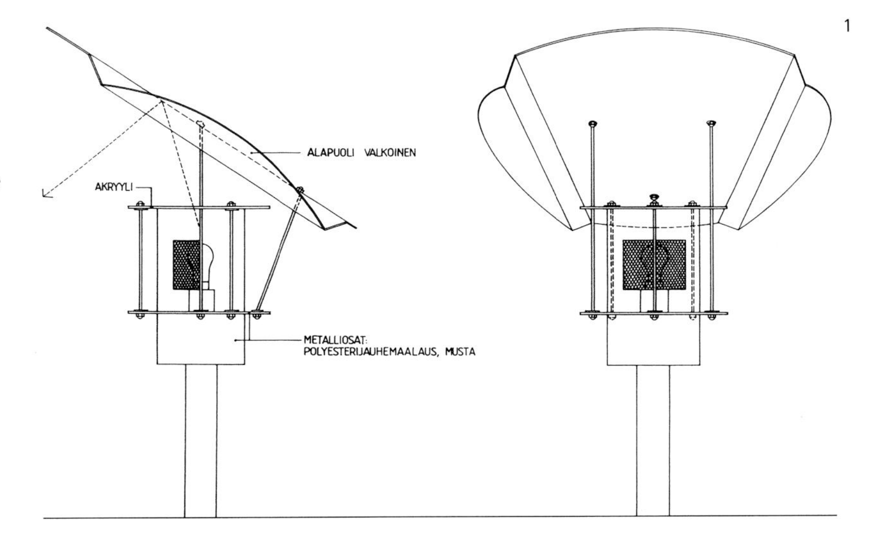

1

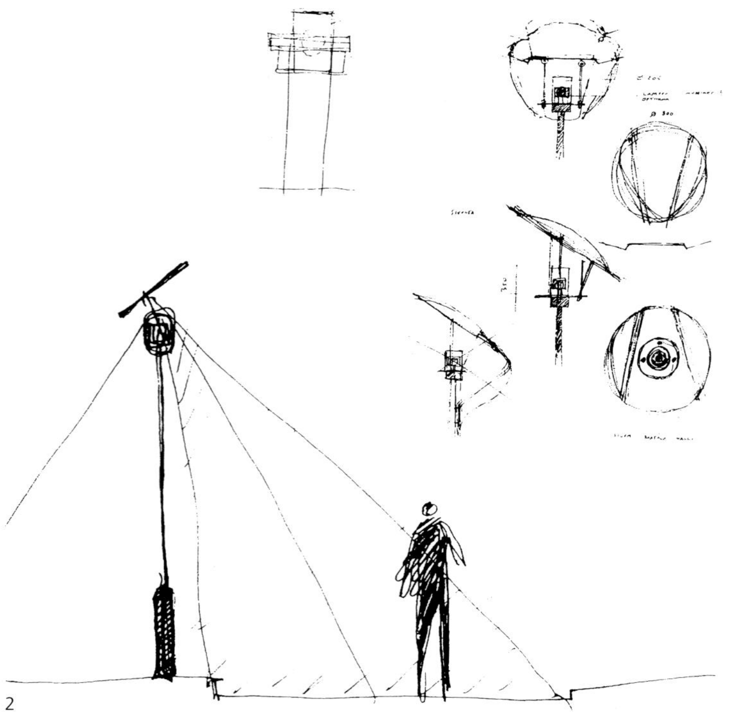

2

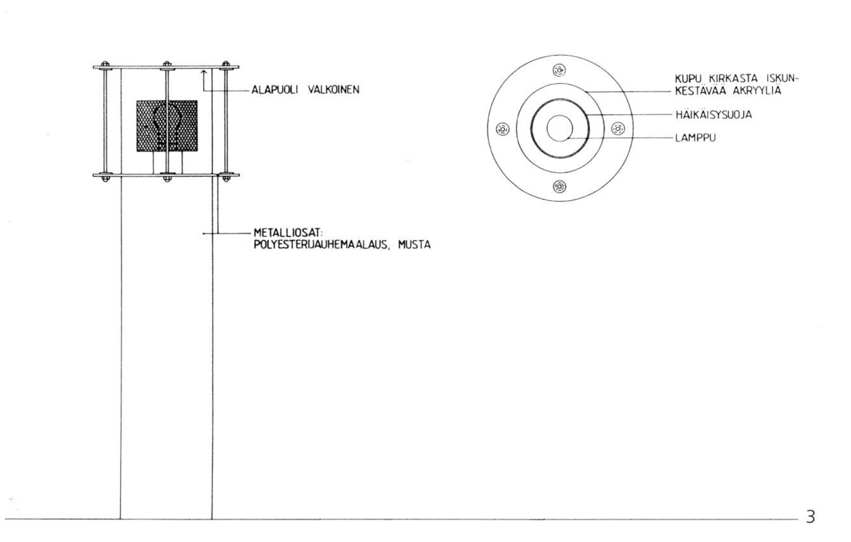

3

1 线图
2 早期草图
3 线图
4 主要通道夜景
5 低照度装置
6 细部
7 细部

1 Drawings
2 Early sketches
3 Drawings
4 Main aisle at night
5 Low lighting fixtures
6 Detail
7 Detail

4

5

6

7

标识系统设计　Sign system design

赫尔辛基地铁标识系统 1970～1983

Helsinki Metro Sign System 1970~1983

1 赫尔辛基地铁标识
2 标志
3 列车内部
4 入口标志及地图
5 入口标志
6 标志
7 剖面、地下车站
8 剖面、地下车站
9 候车平台区的标志

1 Symbols for Helsinki Metro
2 Signs
3 Interior of the train
4 Entrance signs and maps
5 Entrance signs
6 Signs
7 Section, underground station
8 Section, above ground station
9 View of the platform area signs

建立地铁标识系统的目的是为乘坐地铁的乘客提供信息。标识用来指示车站的位置、从车站往城市各个地区去的其它公交联系方式、以及车站地区的方位和其它信息。使用统一的基本要素、清楚的标识和正确的指引就能为地铁服务提供相关的全面信息，创造出一个安全的旅行环境。标识同时也为车站的景观做出了很大的贡献。

标识系统的基本要素是字体、颜色、符号、地图、路线简图和指示板。

乘客需要下列的信息：

——在街道平面：车站的符号、站名、地图、时刻表和车站服务向导。

——在售票大厅：进站标识、出口标识、转乘其它交通方式的指示、地图、时刻表、票务室和车费信息。

——站台位置：站名、出口标识、警告和禁止标识、地图、时刻表、指示板和路线信息。

通过灯光标识系统，乘客获得关于地铁的使用、站台位置、出口路线和转乘服务等重要信息。

对标识的制作质量有非常严格的要求。这不但影响到外观细节的表现，也影响到它们的耐久性及运作性能。标识的结构和盒体都采用不锈钢，面板则是耐用的聚碳酸酯。标识从内部照明，具有很好的鲜明性和易读性。

The purpose of the Metro sign system was to provide information for the passengers using the Metro. The signs are to indicate station locations, together with connecting services to other parts of the city, and give directions and other information within the station area. By consistent use of the same basic elements well-placed, clear unambiguous signs, it is possible to create a safe travelling environment providing comprehensive information on all Metro facilities. The signs also make an important contribution to the appearance of the station.

The basic elements of the sign system are the letter-set, colours, symbols, maps, route diagrams and indicator panels.

Passengers need the following information:

- the street level: station symbol, station name, maps, operating hours and guide to station services,
- in the ticket hall: directions to the trains, exit signs, directions to other forms of transport, maps, timetables, ticket office anf information on fares,
- on the platforms: station name, exit signs, warning and prohibition signs, maps, timetables, indicator panels and route information.

Passengers are provided with important information on using the Metro, platform location, exit routes and connecting services by means of illuminated signs.

Stringent demands were placed on the quality of the signs. This applies to the details of appearance as well as to durability and operation. The frames and casing of the signs are stainless steel, and the front panels are durable polycarbonate. The signs are illuminated from the inside for better visibility and legibility.

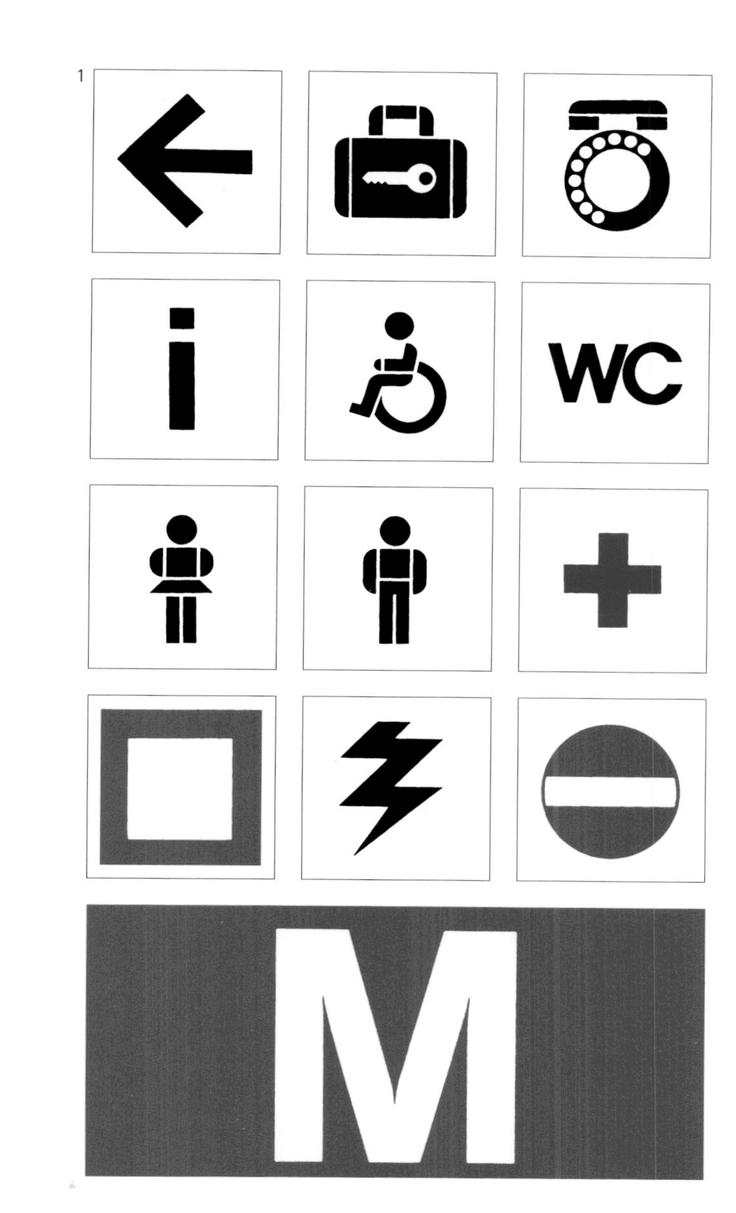

2

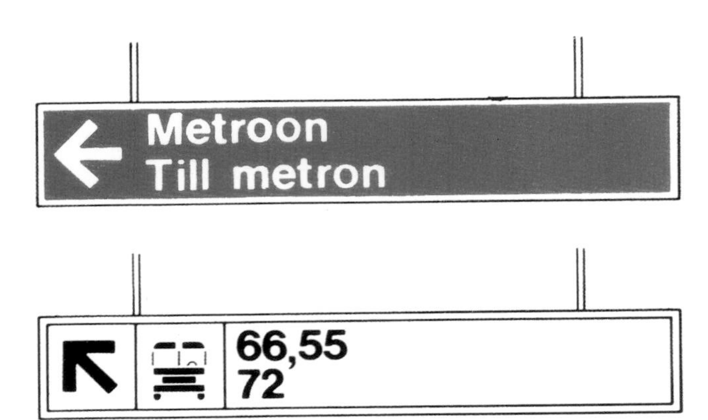
Metroon
Till metron
66,55
72

3

4

Mainos
Mainos
Mainos
M
Hakaniemi
Hagnas

5

M
M
Hakaniemi
Hagnäs

6

9

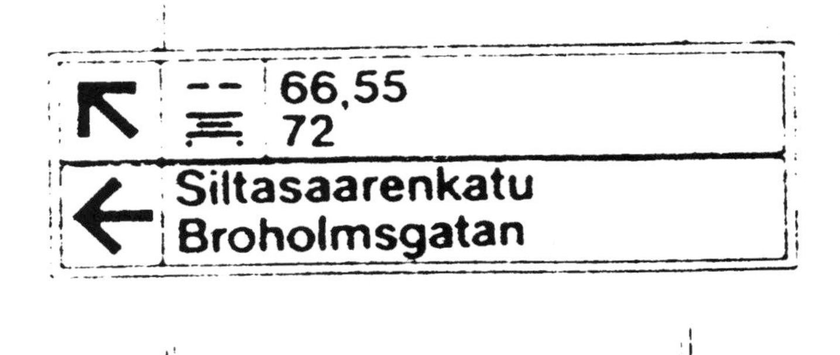
66,55
72
Siltasaarenkatu
Broholmsgatan

Hätätie
Nödutgang

7

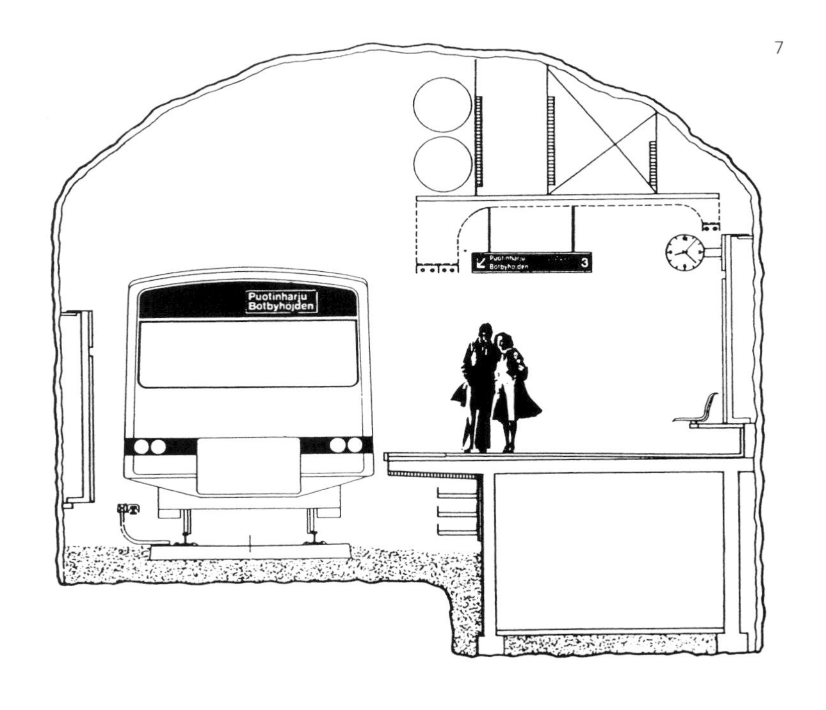
Puotinharju
Botbyhojden

8

Kulosaari
Brändö

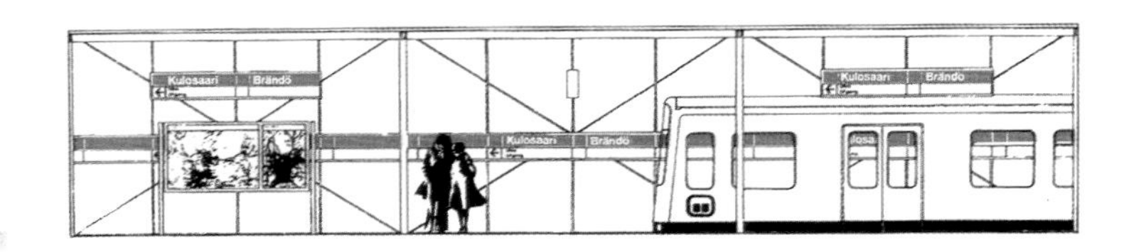
Kulosaari
Brändö

Hakaniemi
Toinen linja
Andra linjen
Hagnäs
Hakaniemen tori
Hagnäs torg
Rautatientori
Järnvägstorget
Hagnäs

芬兰国家铁路乘客标识系统 1976～1983

Finnish State Railways Passenger Sign System 1976~1983

芬兰国家铁路乘客标识系统计划为的是提高乘客的舒适感和安全感，使旅行更畅通。

过去系统缺乏统一：信息公告板的状态非常差，位置也有问题，甚至遗漏一些最重要的点。总之，感官上的统一这个被视为公共交通系统中最为重要的因素已经完全被忽视了。

标识系统的设计工作是从位于Martinlaakso线路上的车站的改造开始的。设计完成后是一个原型评估阶段。在这一阶段，人们获得材料耐久性、抗腐蚀性、人为破坏、实际效果等方面的反馈信息，这一设计步骤非常有效，图尔库、坦佩雷和Lahti车站售票大厅的信息公告板就是按照同样的原则设计的。

根据原型评估阶段的反馈信息，芬兰国家铁路局可以对信息公告板和系统作进一步发展，以形成全国性的统一标准措施。

大多数车站已经根据这个标准建立起标识系统。就连赫尔辛基火车站（由沙里宁建筑师设计）也已经根据这个标准系统做了更新。

The plan for Finnish State Railways' passenger sign system is part of the work carried out to improve passenger satisfaction and safety and make for smoother travel.

The system has been lacking uniformity: information boards have been in bad condition and badly situated, or even missing in some of the most important spots. Above all, a sense of unity, which is now considered of vital importance in public transport systems, has been missing.

The planning work for the sign system began initially with a plan to meet the needs of the stations on Martinlaakso line. The plans were followed by a prototype stage. This provided feedback on the durability of the materials, corosion, vandalism, practical qualities etc., which was useful in follow-up planning. The sign boards in the ticket offices of Turku, Tampere and Lahti stations were planned along the same lines.

With the feedback from the prototype stage Finnish State Railways was able to use the sign boards and system applications to develop their own standard directives for the whole country.

Signs with this system have already been erected at most stations. Also the signs at Helsinki Railway Station (architect Eliel Saarinen) were updated according to this system.

2

1 入口标志
2 标志
3 标志

1 Entrance sign
2 Sign
3 Signs

3

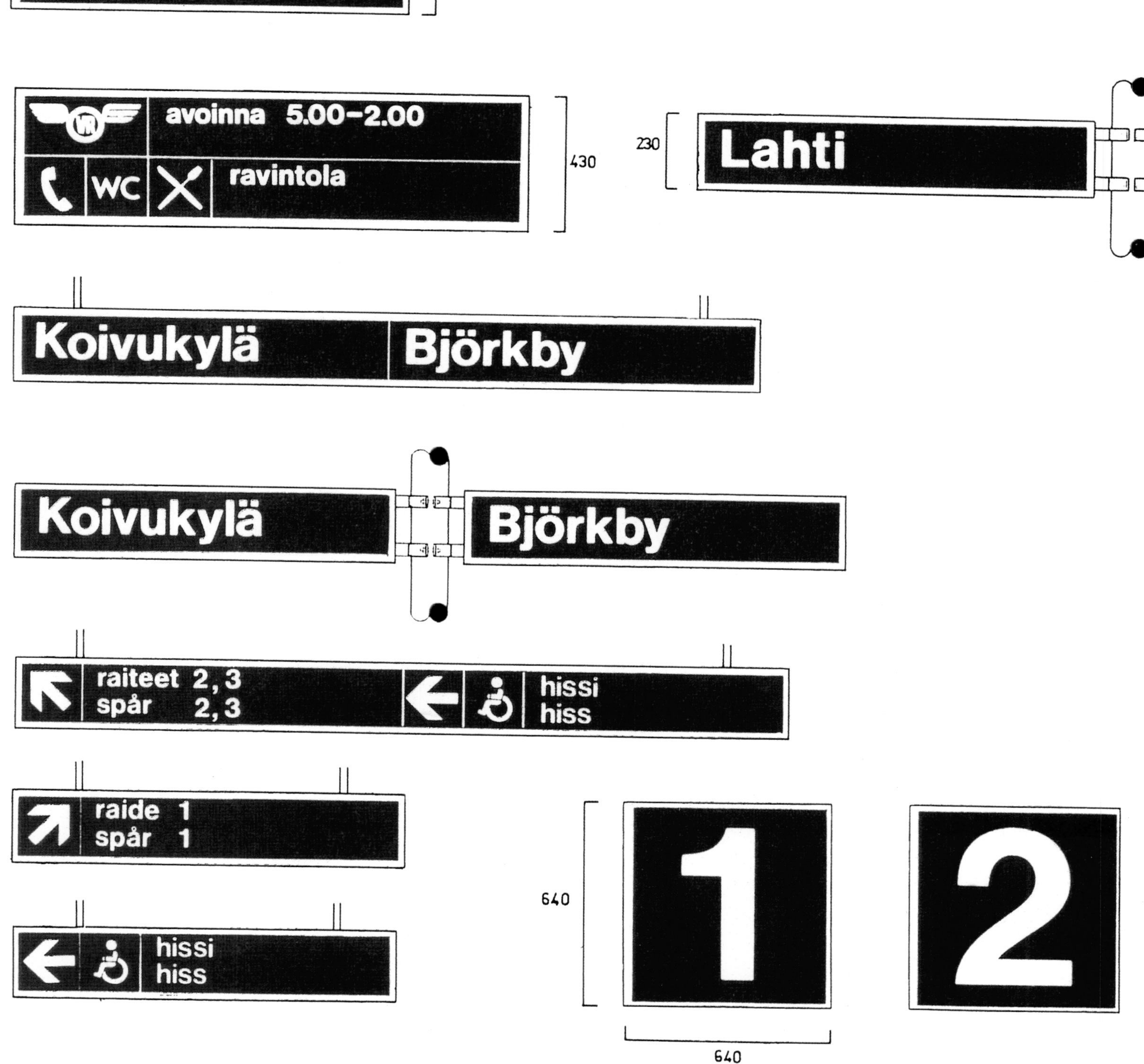

Lahti
500
330
Lahti
avoinna 5.00–2.00
WC
ravintola
430
230
Lahti
Koivukylä
Björkby
Koivukylä
Björkby
raiteet 2, 3
spår 2, 3
hissi
hiss
raide 1
spår 1
hissi
hiss
640
1
2
640

4 标志
5 候车平台区
6 标识

4 Signs
5 View of the platform area
6 Symbols

4

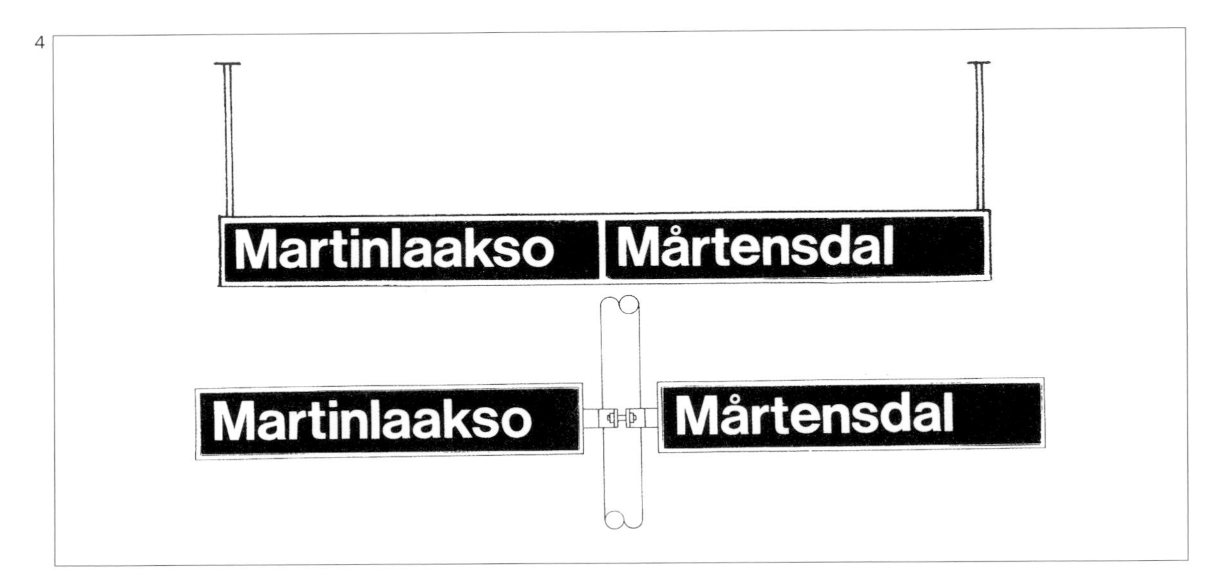

5

1
2
3
R
Reservation
4
5
100
50
5·
6
7
8
9
10
11
12
13
14
WC
15
16
17
R
Reservation
18
19
20
21
22
23
24
25
1
2
26
27
28
29
30
31
32
33
34
35
36
37
38
39
40
41
42
43
44
45
46
47
48

赫尔辛基中央交通站
标识系统
1995

Helsinki Centre
Travel Terminal
Sign System
1995

　　这项工作的出发点是解决赫尔辛基市区内公共交通站点间的乘客引导问题。其重点是解决当前问题比较突出的站与站之间的交通疏导。

　　我们依据乘客的旅行频率把他们分成三组，发现每一个组的人所需要的旅行信息各不相同。不住在赫尔辛基的人和不经常旅行的乘客特别需要获得关于如何到达下一站的清楚的引导。

　　分析了人们在不同交通方式之间的互换行为之后，我们设计了使站与站间尽可能简单的特别路线。随着工程的进展，这条路线成了车站之间的信息通道。设计使这个信息通道从周边环境中凸显出来，这样乘客可以很容易地发现它并安全地顺着它的指示到达另一个目标。

　　赫尔辛基的市区公共交通站点采用了由三个不同层次组成的引导系统：人工信息点、无人信息点和信息塔。人工信息点位于各个公共交通站的操作控制中心，它为乘客提供所有的服务。在无人信息点，先进的技术取代了人工向导，如自动路线计划图表。在这个情况下，信息塔的数量可以减少到7个。放置信息塔的主要原则是在车站之间以最小的合理距离安装，这样即使是第一次来访的乘客也可以跟随信息塔安全地从一个站到另一个站。

The starting point for the work was to find a solution for passenger guidance in the public transport terminals downtown Helsinki. Main attention was paid to the traffic between the terminals,which had been presenting problems.

When the passengers were divided into three groups according to their travelling frequency, it was noticed that the groups needed totally different travel information. The passengers who do not live in Helsinki and are not used to travelling there need clear guidance from one mode of transport to another in particular.

After analysing moving from one mode of transport to another, special routes were planned to make travelling between the terminals as easy as possible.

As the work progressed, this routing became known as the information passage of the terminals. The information passage is to be distinguished so clearly from the surroundings, that passengers can find it easily and follow it safely from one sign to another.

A guidance system consisting of three different levels was adopted in the public transport terminals in downtown Helsinki: manned information points, unmanned information points and information pylons. The manned information points are located in the facilities of the various public transport operators and provide full service to passengers. In the unmanned information points, state-of-the-art technology is used instead of personal guidance e.g. automatic route planners. At this stage, the number of pylons was reduced to seven. The main principle of the pylons was to place them at least at reasonable distances from one another so that even an unaccustomed passenger can move safely from one terminal to another by following the pylons.

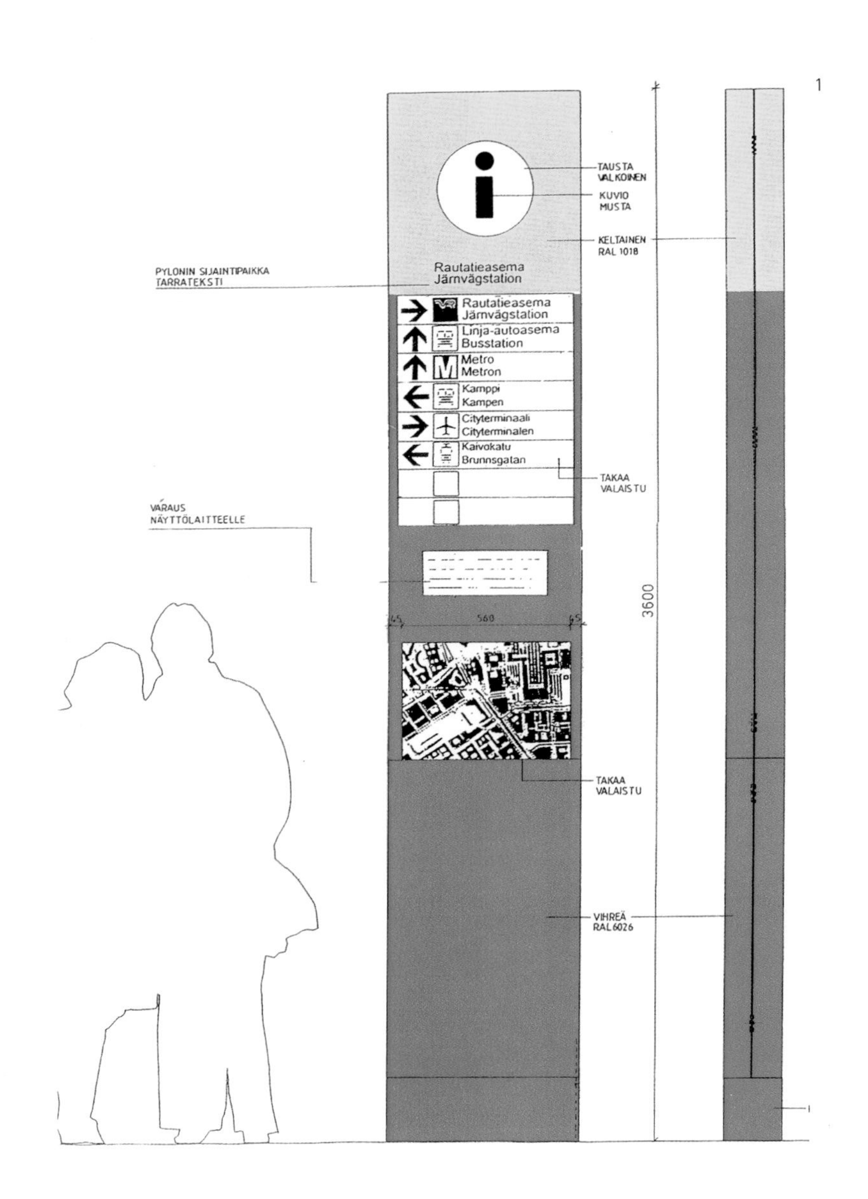

2

3

1 标志塔图
2 街景中的标志塔
3 移动式标志塔控制下的标志塔区域
4 位于赫尔辛基车站前方的标志塔

1 Pylon drawings
2 Pylon in streetscape
3 Pylon sites were checked by moving pylons
4 Pylon in front of Helsinki Railway Station

4

i
Rautatieasema
Järnvägsstationen
2, 3B, 3T, 6
M
Rautatientori
Järnvägstorget
2
Katajanokka
Skatudden
3T
Eteläsatama
Södra hamnen
Rautatientori
Järnvägstorget
Katajanokka 1,7 km
Skatudden
Eteläsatama 1,9 km
Södra hamnen

平面设计 Graphic design

1

2

早在学生的时期，平面设计作为一种爱好已经成为我工作的一部分。作为一个建筑学的学生，我们不得不面对许多根本实现不了的设计。而在平面设计中，作品却可以即刻实现。设计的实现给以设计为业的人带来极其珍贵的满足感。

与进度慢、周期长的建筑设计工作相比，平面设计无疑提供了一个很好的调剂。同时，它提出的是与建筑不同的微型的功能问题。印刷艺术的快速与建造艺术的缓慢结合形成很好的搭配。

我们的平面设计作品包括：企业形象、版式设计、书籍装帧、海报和商标设计。

My working in the field of graphic design started as a hobby while still student.
A student of architecture is obliged to see many unrealised plans during the studies. In the graphic design I found a field whose plans were realised immeadiately. All this lead to the fulfilling experience of having one's plans realised which is of vital importance to the designer.
Graphic design offers satisfactory change to the more slow and long-term work of the architect. At the same time it offers a functioning miniature scale usually not to be found in architecture. The typical punctuality of printing art is a pleasant counterbalance to the altitude of building.
Here are some examples of the printed matter as a result of graphic design: corporate images, typography and lay-outs, book design, posters and trademarks.

3

4

1 Tekniikan Suomi maailmankartalle 海报
与 Esko Miettinen 合作
2 Taidetapahtuma 海报
与 Timo Paasimaa 合作
3 Otaniemi Finland 海报
4 Kesähotelli 海报
与 Ola Laiho 合作
5 Ikituuri 标识
6 《体育与休闲》封面
7 《Erik Bryggman》封面
8 Marli 标志

1 Tekniikan Suomi maailmankartalle poster
in collaboration with Esko Miettinen
2 Taidetapahtuma 1 poster
in collaboration with Timo Paasimaa
3 Otaniemi Finland poster
4 Kesähotelli poster
in collaboration with Ola Laiho
5 Ikituuri symbol
6 Sports&Leisure book cover
7 Erik Bryggman book cover
8 Marli symbol

5

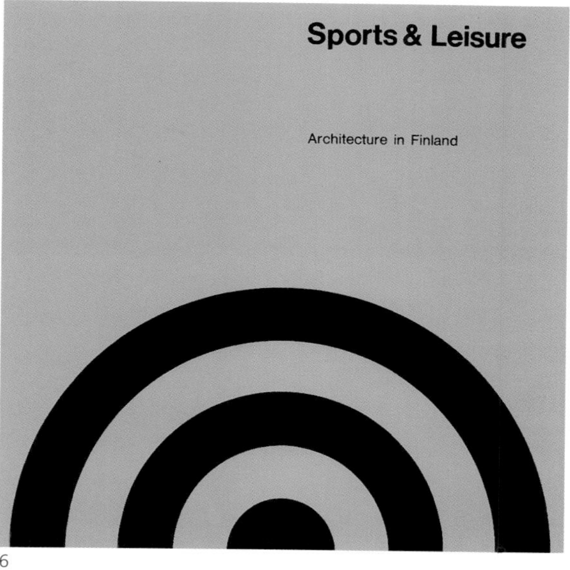

6

7

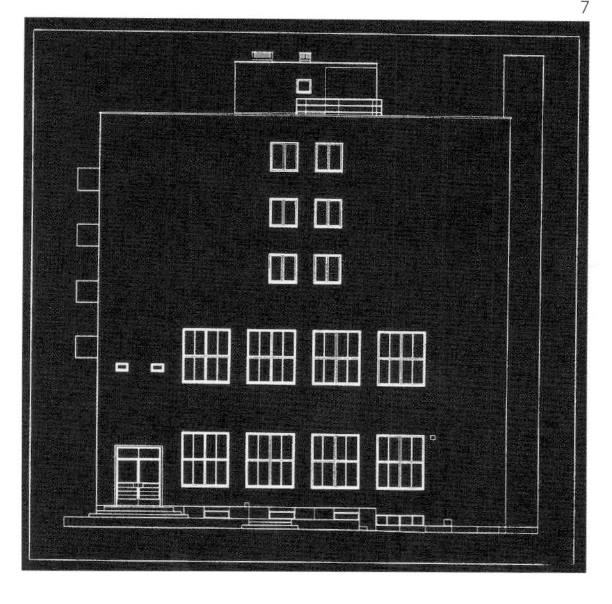

8

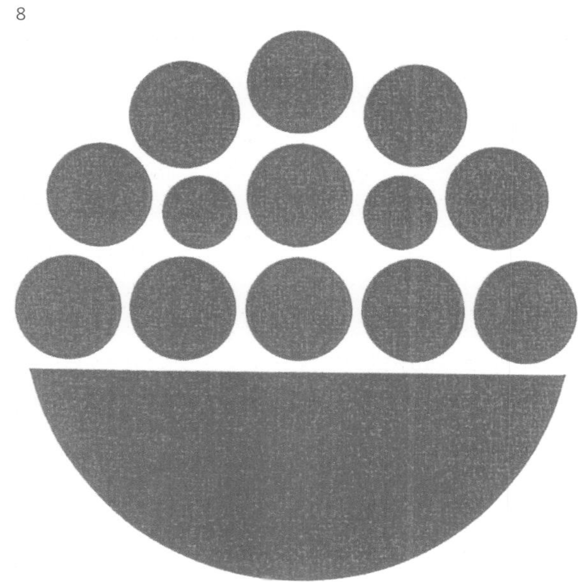

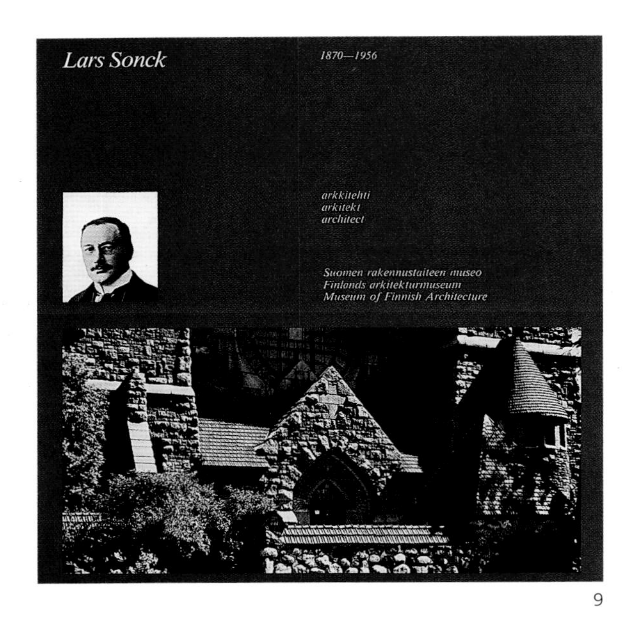

9

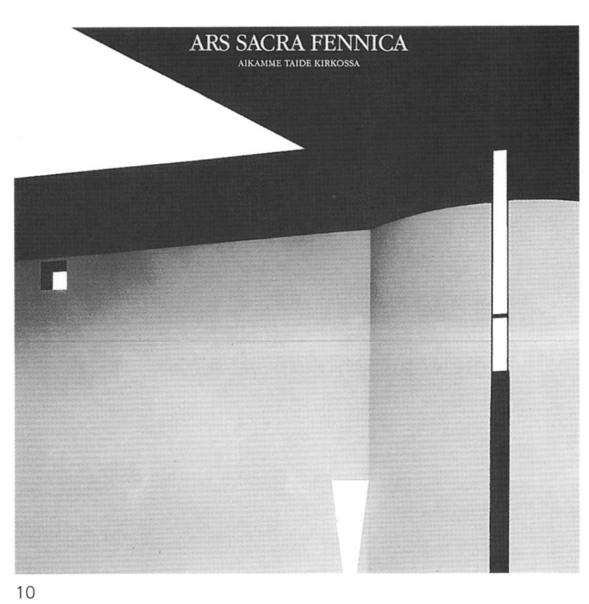

10

11

9 《Lars Sonck》展览邀请卡
10 《Ars Sacra Fennica》封面
与 Jyrki Nieminen 合作
11 Novaco 商标
与 Jyrki Nieminen 合作

9 Lars Sonck invitation card
10 Ars Sacra Fennica book cover in collaboration with Jyrki Nieminen
11 Novaco trademark in collaboration with Jyrki Nieminen

论文 Articles

检测人类对环境的反应

Measuring human responses to the environment

摘要

随着试验技术的发展，检测和分析人类对环境的反应成为一个复杂而艰巨的任务，它包含了许多因素。这种复杂性要求使用一系列的研究工具和方法。因此，这项研究被认为是把客观的研究联系于城市设计的一次试验。在心理学与形象化分析方面的基础研究形成了这项研究的技术基础。在相关领域里，Lynch，Osgood，Thiel，Winkel和许多其它的科学家们所作的工作对这项研究产生了积极影响。

这项研究的客观性意义在于关于人们对城市环境的反应方式的深层思考。这项研究基于这样一个前提，即通过人们对空间的反应的某些理解，以此来获得描述和定义空间的更加完整的理解。

为了得出人们对环境的反应，选取了一组建筑系的学生和一组非建筑系的学生，以进行对比。他们被要求对幻灯片上的街道风貌给出一个合理的评价。

这样，在基于Kasmar的关于检测城市环境主要因素的重要反应的研究之上，进行了一次图片评价调查。这种方法考虑了语义上的差异，它使用完全相反的形容词来得到对一个物体的隐含印象。

最后，用简单统计方法对调查结果进行评价和分析。

人们对环境条件的反应是试图研究人们对空间好恶方面的起点。研究英国的一个小镇，然后把它的原理应用到美国的一个小镇的设计计划中，这样好像不是一种有效的途径。这样一种研究方式没有考虑到也许存在于两个国家和人们之间的社会、文化以及心理方面的差异。根据Rapoport所说，文化背景的差异将导致人们在感觉和欣赏水平上的差异。为了针对这些不同背景的人们而进行成功和完整的设计，设计者必须对存在于这些水平上的差异有一个清醒的认识。

同理，在觉察熟悉的和不熟悉的环境方面也存在着差异。一个对环境（假设是人造的或模拟的）不熟悉的人不大可能对实际的空间作出相应的反应。Hall在他的著作《隐藏的空间》中涉及到这个问题。他说我们的城市产生了很多不同类型的人群，例如在贫民窟里、精神病院里、监狱里和郊区里等，每

ABSTRACT

The experimental development of techniques for measuring and analyzing human responses to the environment is a complicated task involving many factors. This complexity requires using a variety of research, tools and methods. Because of this, the study should be regarded as an experiment in relating objective research to urban design. Basic research in psychology and visual analysis have produced the basis of techniques used in this study. Work in related fields which has undertaken by Lynch, Osgood, Thiel, Winkel and many others underlies many elements of this study.

The objective of this study is to give some insight into the way people respond to urban environments. The study is based upon the premise that with some understanding of the response of people to space, a more complete understanding of the terms used to describe and define space may be gained.

In order to elicit response to the environment, a group of architectural students were selected for comparison with a group of non-architects. These groups were asked to evaluate street views on slides.

An image evaluation questionnaire was developed based on Kasmar's research to measure the critical responses to the character elements of urban environments. This method incorporated a semantic differential, which uses polaropposite adjectives to obtain connotative implications of an object.

The results were evaluated and analyzed with simple statistical methods.

Responses of people to environmental conditions should be the starting point in an attempt to study the likes and dislikes of people with respect to space. To study the English townscape and apply the principles that exist to the design of a small town plan in the United States, is not necessarily a valid approach. Such a study would not take into account the social, cultural, and psychological differences that may exist between the two countries and peoples. According to Rapoport, the differences in cultural background, lead to differing levels of awareness and appreciation in the cross-section of population being designed for. To design succesfully and adequately for these differing backgrounds, the designer must have some index of the differences that exist on these levels.

There is also a difference in perceiving familiar and unfamiliar environments. It is not likely that the person who is not familiar with the environment (presented artifically or simulated) will react or respond the same as he would to the actual space. Hall touches this problem in his book The Hidden Dimension. He states that our cities are creating different types of people in their slums, mental hospitals, prisons, and suburbs and each world has its own set of sensory inputs, so that what crowds people of one culture

一个世界都有它自己的一套感官输入，因此在一种文化背景下适应的规律不一定适应于另外一种文化。与此类似，一种也许对某人产生压力的侵犯行为对另外一个人来说也许是中性的。接着他还说，“我们城市的重建工作必须建立在调查研究人们的需求之上，以及懂得居住在城市里的各种人群的不同感官世界。”“人类与环境的联系就是人类感觉器官的功能加上他的器官如何调节以此来对环境作出反应。”

Mckechnie对环境反应倾向的评价做了一个环境反应详细目录（ERI）。ERI的合理性在于它从人性的心理方面得出结论：人们以一种不变的有特点的方式对待每天的物理环境，就像他们对自己或对别人的那样重复的闲聊。他研究了人们的休闲娱乐方式，发现了几种因素，以此来描述人们对环境的反应。

根据Mckechnie所讲，很明显研究环境心理必须要考虑人们的环境倾向，即态度、信仰、价值观和情感的综合体现，需要研究人们的环境行为，以及预测他们下一步的环境反应。

这项研究的结果表明在察觉城市环境方面，建筑学者和非建筑学者之间是有差异的。建筑学者总体来说似乎比非建筑学者在感觉城市环境方面更具有情感。他们也似乎比没有学过建筑的人更能很好地描述空间的特征。因此，我认为这是因为他们的教育背景不同的原因。

这项研究也清楚地表明熟悉的城市环境（例如Raleigh 的Hillsborough 街）在感觉上比较相似，而不熟悉的环境则不一样。从这些反应的结果可以表明，这项研究作为一种方法在决定城市环境特征的重要性和意义方面有一定的潜力。

问题

我们知道在环境和人类行为之间有某种联系。但是，我们不知道那种联系的关键变量是什么。

符号系统在考察存在的环境方面有它自己的意义，例如Lynch 和Thiel建立的符号系统。如果对每一个符号体系都有非常深刻和完整的理解，人们也许能把它们组合成一个设计工具，但是对大多数而言，它们太难了，以至不

does not necessarily crowd another. Similarly, an act that releases aggression and would therefore be stressful to one people may be neutral to the next. Later, he says, "The rebuilding of our cities must be based upon research which leads to an understanding of man's need, and a knowledge of the many sensory worlds of the differing groups of people who inhabit cities." "Man's relationship to his environment is a function of his sensory apparatus plus how this apparatus is conditioned to respond."

McKechnie has developed the Environmental Response Inventory (ERI) for the assessment of environmental dispositions. The rationale underlying the development of the ERI is drawn from the psychology of personality: that people relate to the everyday physical environment in stable, characteristic ways, just as they relate to themselves and to others according to enduring patters. He has studied man's leisure recreation behavior and found several factors which describe man's response to environment.

According to him it is also clear that research in environmental psychology must take into consideration the environmental dispositions – the configurations of attitudes, beliefs, values, and sentiments – of the people whose environmental behavior are being studied and whose future environmental responses are being predicted.

As a result this study shows that there is a difference in perceiving urban environments by architects and non-architects. Architects in generally seem to be more emotional in perceiving than non-architects. They also seem to have a better ability of describing the character of spaces than the non-architects. This I think is because of their educational background.

The study also shows quite clearly that the familar urban environments (e.g. Hillsborough Street in Raleigh) are perceived more similarly than the unfamiliar. The results obtained from the responses shows that the study has potential as a methodology for determining the significance and meaning of the character of urban environments.

PROBLEM

We know that there is a relationship between environment and human behavior. But we do not know yet what are the critical variables of that relationship.

Notation systems such as those developed by Lynch and Thiel have their own value to the observation of existing environments, and with a very deep and complete understanding of the mechanics of each notation systems, one might be able to incorporate them as a designing tool, but for the most part they are too difficult to achieve this level of understanding for incorporation into the design method. They find better application when used for the observation of

能达到能组合成设计方法那样的理解深度。人们发现它们可以很好地应用于已存在的空间和时间，对新产生的设计就显得力不从心了。

其它的研究还考察了空间质量，定义了一整套组成微观或宏观环境的因素，从结构到实际物体，从抽象的体积到大量建筑群之间的联系。Cullen的理念是借助于理解这些因素，我们能够深切地理解组成空间特性的定义。深刻地理解这些特性，并使用那些有利的因素，这样将有助于更好的设计。

第三个领域是模拟技术，在这个领域里，人们试图定义环境的质量，特别是试图感觉一些建筑质量的可预测性。

所有的这些方法都与对环境的反应这个大的信息主体有一定的联系，当然，所有的这些都停留在思维摸索上。没有哪一个达到了精确和可预测的程度，从而能实际地帮助建筑学家们改进他们对人类环境的设计。在这个领域里缺少的信息是调查人们对组成环境的建筑的反应。依靠心理学家和社会学家的关于人们对可视的和物理的刺激物的反应的研究成果，建筑学家们不能真正地了解用户对他们实现的作品的反应。因此，对城市空间质量的主观上的理解是直觉的，要达到更深一层次的理解的途径就是消化吸收这些领域里的优秀作品。一些深刻理解空间质量的先驱们的作品是Lynch，Rasmussen，Norberg-Schulz，Hall和少数其它的一些人的作品。方法学上的先驱们的作品是例如Sommer，Osgood，Winkel，Rapoport，Craik，Canter，Mckechnie，Sanoff 以及一些其他人的作品。

Cullen建立了一系列城市元素特征的目录。这种思想很大程度上依赖于为理解那些特征而进行的形象的分析，并且他确实深刻地完成了理解的任务。特性的表征词汇的建立有助于把这些特性应用到后来的建筑和城市空间设计中去。这些特征是建筑学家们建立和定义的，然而它有可能与这些设计空间的使用者的反应特征不相符。看起来好像建筑学家和设计师们有他们自己的设计语言，而这种语言不一定被那些没有接触过建筑理论或实践的人们所理解。

在对环境刺激物的反应的这个领域内，绝大多数的工作建立在一些相对模糊的要素的理解之上，我们认为这些要素构成了建筑。Norberg-Schulz说人们总是以一种满足于现状的态度来看待周围的环境。他告诫我们必须正确理

existing spaces and sequences than for the generation of new designs.

Other studies have looked at spatial quality, and have defined a whole range of elements which make up the micro-or macro-environment, from texture to physical objects, from abstract volumes to relationship between masses of buildings. The idea of Cullen is that by understanding the elements, we develop insight into the definition that might be make up spatial character. Deeper understanding of the character will lead to a better ability to design, using the positive ones.

A third area in which there has been some attempt to define the quality of the environment, and, particularly, to affect some degree of predictability of architectural quality is the area of simulation techniques.

All these methods of approach have some relationship to the larger body of information concerning the response to the environment, and certainly all are the product of inquiring minds. But none have reached the degree of accurancy and predictability which will really help architects to improve their designs for the environment of man. The indicated lack of information in this involvement area is in the area of the investigation of the response of humans to the elements of architecture that constitute the immediate environment. Expecting the works by psychologists and sociologists in human response to visual and physical stimulus, architects have no real understanding of the reactions of the user to the works they implement. The understanding of the subjective qualities of urban spaces has therefore been intuitive in nature, and, the approach to greater understanding has been that assimilation of the reading of a few selected works in the area. Some of the progenitors of deeper understanding have been the works of Lynch, Rasmussen, Norberg-Schulz, Hall, and a few others. Progenitors of methodology have been the works of Sommer, Osgood, Winkel, Rapoport, Craik, Canter, McKechnie, Sanoff and a few others.

Cullen has developed a fine catalogue of the character of urban elements. The ideas are largely dependent upon the use of visual analysis to understand character, and the extent to which he accomplishes the aims of understanding are indeed great. The development of a vocubulary of characteristics aids the application of such characteristics to the subsequent design of buildings and urban spaces. These characteristics are developed and defined by architects for the tastes of architects, however, and do not, as a role, pertain to the reactions character causes in the people who are users of the designed spaces. It would seem that architects and planners have their own language of design, and that this language is not necessarily understandable to people who are not connected with architectural theory or practice.

解和判断事物，使它们更好地为我们服务。我们判断一种事物具有它们体现出来的某些表现或性质，这些性质使我们产生一种积极的或消极的反应。我们也对所有的事物都有着不同的态度，那种态度也由我们所处的形势所决定。

假设

人们对城市环境的看法是如此的不同，以至于依靠于某一个人的关于什么组成了特定的主观空间质量的判断是错误的。为了测出某人对环境刺激物的反应，他必须要置身于特定的背景下，他可以用以下的一个或几个条件所限定：年龄，性别，文化，教育等等。几个群体将被互相比较，以得出存在于他们之间的觉察差异以及对城市刺激物的反应的不同。

——拥有相似的工作或受教育背景的人群其对给定的环境刺激的反应相似。

——假定相似的刺激条件，在不同的受教育背景下的人群，他们的反应有很大的不同。

选定这些假设，从而希望得出一个成功的方法，从这些方法上再启动这个领域内的更进一步的研究。

空间的选择

从作者拍下的300多张幻灯片中，选取一些较好的模拟的幻灯片。复杂的以及包含很多城市环境特征量的幻灯片优先选取。

群体的选择

在测试中选取了以下的群体：

1. 从建筑系毕业的学生和从政治学毕业的学生（NCSU）；

2. 一年级和二年级的建筑系的学生（HTKK）。

调查

在检测人们对建筑或环境的反应时，或者测试人们对关于空间觉察的某些概念的印象时，特意使用了语义上的等级差异。从而能够对概念和印象或

Most of the work in this area of response to environmental stimuli are based upon the intuitive understanding of a few relatively obscure elements which, we say, make architecture. Norberg-Schulz shows that we are highly dependent upon seeing our surroundings in a satisfactory manner. He says that we must understand or judge things to make them serviceable to us. We judge things be certain manifestations or properties which they posses, and these properties cause us to react in a positive or negative manner. We also have differing attitudes toward all things, and that attitude is dictated by the situation we are in.

HYPOTHESES

People view urban environment so differently that to depend upon the judgement of one person as to what constitutes a particular subjective quality of space would be erroneous. In order to index the response of the individual to this stimulus, he must be placed within a specific ground which may be defined by one or more of the following controls: age, sex, culture, education, etc. The groups defined will be compared to determine the differences that might exist in their perception, and response to the urban stimulus.

– Individuals within similar occupational or educational groups will respond similarly to a given environmental stimulus.

– Given similar conditions for stimulus, there will be a significant difference in response between differing educational groups.

The hypotheses have been selected in the hope that the evolution of a succesful methodology will result, from which further intensive study in this area may be initiated.

SPACE SELECTION

The slides used in simulation were selected from among 300 slides photographed by the author. Complex slides were preferred and slides which contained most variables of urban charactersitics.

GROUP SELECTION

There were used the following groups in the tests:

1. graduating achitectural students and graduating political science students (NCSU)

2. first and second year architectural students (HTKK)

QUESTIONNAIRE

Semantic differential is intented for use in measuring peoples' reactions to buildings or environments, or, for measuring the connotations of peoples'

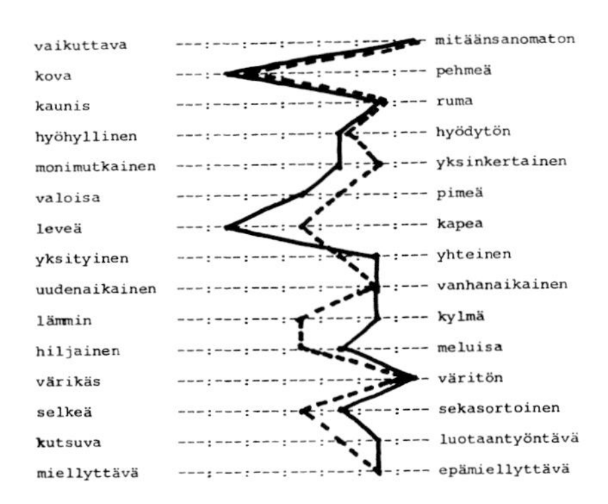

对特定的物理环境有一个清晰和明确的态度。同时，从Kasmar的66个可替换的描述特定的建筑环境的尺度中选取了合适的环境描述尺度。

在调查表中还加入了一些短的个人资料信息。

images of certain concepts relating to space perception. It can clarify and record attitudes toward concepts and images, or toward specific physical environments. Environmental description scales were selected from Kasmar's 66 alternatives suitable for describing specific architectural environment.

A short personal data was added to the questionnaire.

模拟

幻灯片被投影在墙上，每一张5到10秒钟，并且每两张幻灯片之间有相同的时间间隔。在评价语法差异等级时使用这些相同的幻灯片。

SIMULATION

The slides were projected on a wall for five to ten seconds each, with a similar interval between each slide. The same slides were shown for the semantic differential scales.

分析

这次调查的客观性在假设的框架下列出来了。每一个群体之间（列出环境）在感觉上差异被列成图表。分析是对提出来的假设的一次回顾。

这些图表表明在两个群体之间的感觉方面有差异。同时，它还表明学过建筑的人们总体来说对他们的环境比较挑剔。他们也比没有学过建筑的学生更能够很好地描述空间的特征。

图表还表明人们对熟悉的环境的反应相类似，而对相对不熟悉的环境的反应则大大的不同。

从这些反应的结果可以表明，这项研究作为一种方法在决定城市环境特征的重要性和意义方面有一定的潜力。

在下一步的研究中，建议还需要使用更多的复杂的方法以及大量的样本群体。

ANALYSIS

The objectives of the survey have been outlined in the hypothetical framework. The analysis is a review in relationship to the hypotheses proposed. The difference in perceiving between the groups in each (environmental display) is shown on the diagrams.

These diagrams show that there is a difference in perception between the two groups. The diagrams also show that architects in general are more critical of their environment. They seem to have also a better ability to describe the character of spaces than the non-architects.

The diagrams also show that the familiar environmental displays are perceived almost similarly and the greater difference occurs in more unfamiliar displays.

The results obtained from the responses shows that the study has potential as a methodology for determining the significance and meaning of the character of urban environments.

More sophisticated methods should be used and a larger sample groups is needed for future research.

设计标识系统

随着我们生活环境的日益复杂，城市和建筑中的指向设计的难度变得更大。我们设计的巨型综合性建筑、大学和办公楼占地面积巨大，功能复杂多样。由于功能分化越来越细，场所越来越国际化，公共空间的定位和指向的要求就变得越高。为特定的环境设计特定的标识系统并不一定是最有效的做法。这样做不仅昂贵，同时也不能保证标识系统的清晰明了。因此非常有必要形成一套普遍适用的标识系统，能够同时满足城市环境及新旧建筑内部的标识要求。

解译环境

人类的解译能力是有选择性的。在观察的那一片刻，他的动机、经验和知识形成一种过滤器，使得一次只能有一个被观察到的特征通过单通道系统进入中央神经系统，然后到达长效记忆系统和动力中心，这一部分神经系统才能指挥我们的动作。

人的解译能力同样也被单一环境制造的饱和界限所限制。人们会对观察范围中发生的变化作出反应，而这些变化信息的获得和可靠性受到信息集中度和观察时间长短因素的影响。步行者的观察范围普遍大于驾车者的观察范围。如果某个标识需要发挥作用，那它一定要从背景里脱颖而出。也就是说，与这个标识相比，周围环境中的“视觉污染”必须降到足够低的水平。

一个人对环境的感知是这个人与他所处的环境交互作用的结果。环境表现出的是自己的特征，以及特征与特征之间的联系，观察者根据自己的需要将它们选择性地重组，以此确定他所接收到的信息不同的重要程度。

在寻路的行为中，人对环境的印象，以及人主观生成的外界情况的画面起着至关重要的作用。这个印象是直接观察和过往经验两者的结合，它被用来解译信息和指导行动。

70%～80%的信息通过视觉观察传递。经神经系统处理过的环境图像永远不会是百分之百“真实”的。在头脑中经历和认知的东西在尺度和强度方面并不完全与外部物质世界中的实际情况相对应，因为实际物质的刺激与人所接受的图像之间的关系并不稳定。我们的观察力具有极大的选择性。

一个影响观察内容可信度的重要因素，是这个信息在我们观察范围之内存在的时间。如果刺激的持续时间较长，可以补偿中途短暂的间隙。通过读与写出来的文字，分别产生出短效与长效影响。闪烁的光源比持久的光源更引人注意，就是因为它不同于照明用的普通光源。

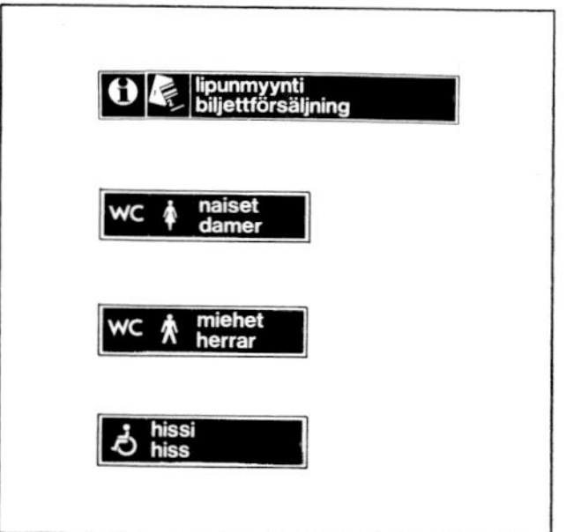

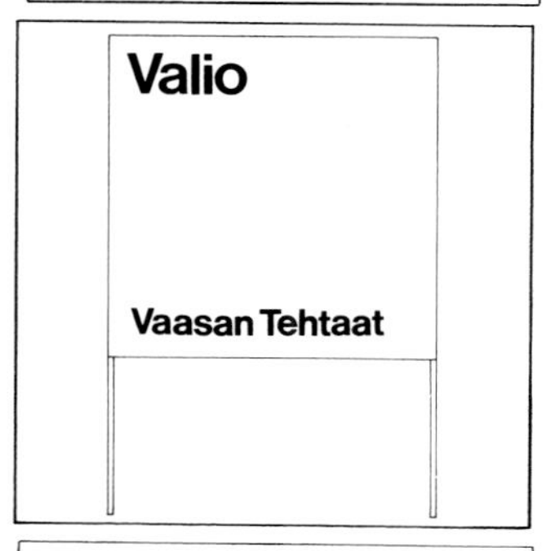

Designing a sign system

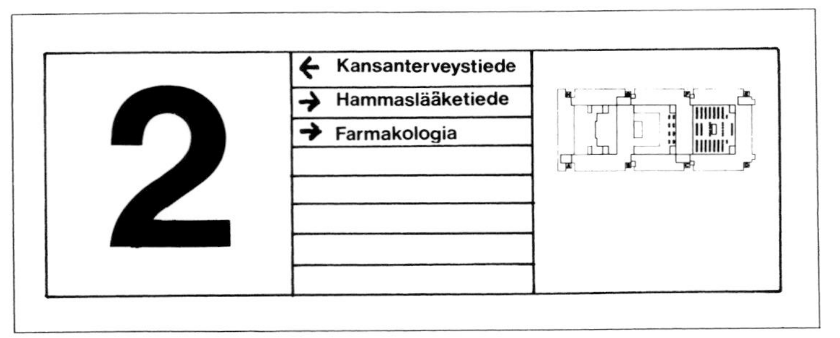

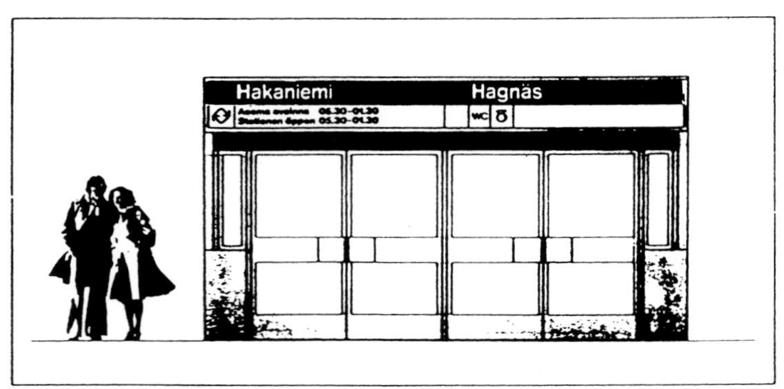

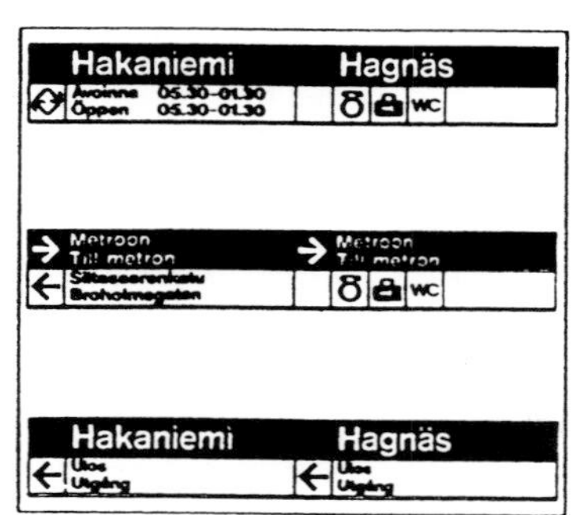

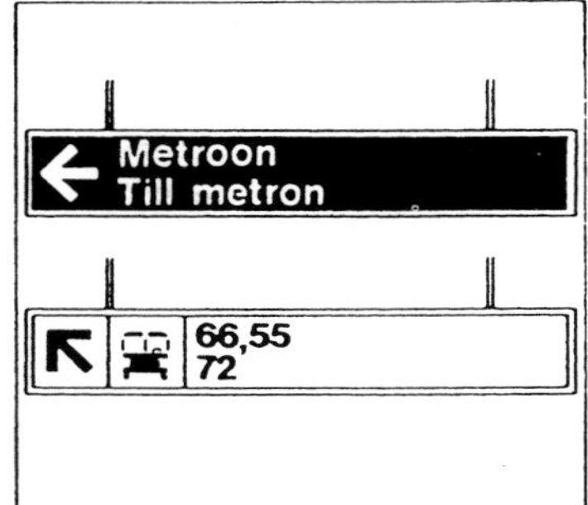

ABCDEFGHIJKL
MNOPQRSTUV
WXYZÅÄÖÇÆ
abcdefghijklmno
pqrstuvwxyzåäö
/&",,!?()-
1234567890

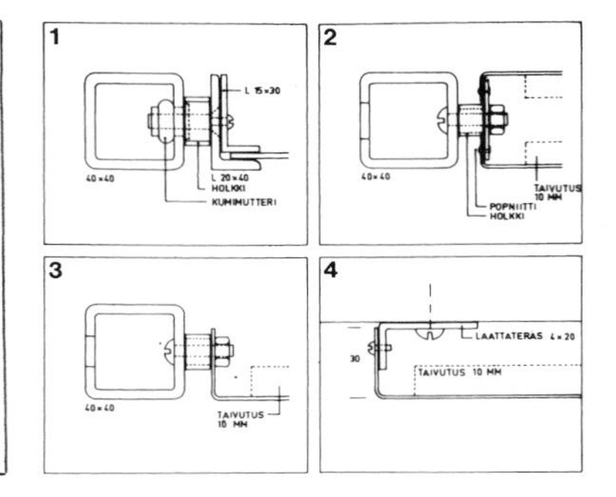

As our environment becomes increasingly complex, orientation in towns and buildings gets more difficult. We build huge complexes, universities and office blocks covering whole areas and containing a wide variety of functions. Guidance for he public is becoming increasingly important as functions diversify and become more international. A sign system designed for one particular situation is not necessarily the best; it is expensive and one cannot always quarantee the clarify of the signs. It is important to develop sign systems of general validity which suit both urban milieux and existing and new buildings.

DECODING THE ENVIRONMENT

Man's decoding ability is selective. His motivation, experience and knowledge at the moment of observation comprise a filter which allows only one observed feature at a time through into the single-channel system to the central nervous system, and on into our long-range memory and motor nerve centre, which decides our movements.

Man's decoding capacity is also affected fundamentally by the saturation level produced by a monotonous environment. We react to changes in our field of observation and the making and reliability of these observations are affected by the degree of concentration and how long the information is observable. The pedestrian's observation field is generally much wider than a driver's. If a given signal is to be 'put over', it must stand out from its background, i.e. the visual "noise level" of the surroundings must be sufficiently low in relation to the signal.

A person's image of his environment is the result of mutual interaction between him and that environment. The environment present features to be distinguished and relations between them, and the observer selects and arranges them to suit his own ends, attaching significance to what he sees.

In finding the way, a person's image of his environment, his generalized picture of the outer physical world, is of vital importance. This image is the product of direct observation and former experience, and is used to interpret information and guide action.

70–80% of all information is transmitted via visual observation. The image gained of the physical environment via the senses is never completely 'real'. Mentally experienced and recognized observations do not correspond directly to the physical system of the outside world in dimensions or intensity because the relationship between the physical stimulus and the experienced observation is not stable. Our observation power is extremely selective.

One feature greatly affecting reliability of observation is how long the information is in our observation field. If the observation stimulus lasts a long time, short breaks can be compensated. The spoken and the written word are good examples of short and long term effect. A winking light is more effective than a continuous light because it stands out from lights used for other purposes.

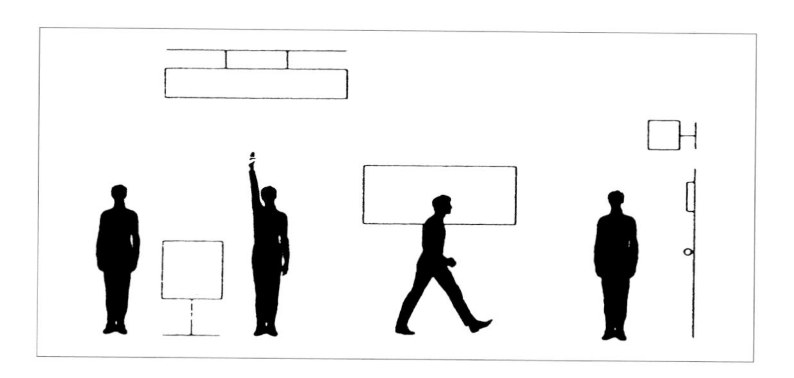

方位

形成功能合理、环境宜人的步行环境的前提是该区域有一个良好的定位指向特征。一个人在这样的环境中始终知道自己身在何处，以及下一个目的地的位置和行走路线，这一点极其重要。主要的定位系统有两种：

1．环境的形状特征，及通过视觉线索获得的场所感和方位感。

2．标识符号提供的方位感。

对于直接的定位，下列因素非常重要：

——清晰的路线。在这方面，方格网平面系统性能较好。

——对角线路导致方位感的弱化。因为人们较难辨认直角以外的路径变化。

——较少的对称有助于确定方位。

——对称与细节有利于估测尺度。

——中央空间的原则有助于方向感的形成。中央大空间（广场、中央大厅等）可以帮助人们把握周边的情况和尺度。

——从远距离多方位可以看到的地标、建筑物、特异的地貌帮助人们确定方向和距离。

——照明。重要的定位点需要得到充分的强调。

交通渠道的极其重要：交汇点、交换站、终端尤其重要（清晰的交通分流至关重要）。

当街道系统布局完善，形式规则，就能方便人们的通行及定位。单调和缺乏层次的街道令人缺乏方向感。这时，标识就显得极为重要。

指向

随着公共交通的日益发达，需要重视那些提供有关公交服务信息的标志系统的设计。

人类环境改造学是标识系统设计的前沿知识。将所有的信息源整理成人们能够方便而清晰无误地接受的形式是设计中至关重要的部分。视觉与听觉的信息在公共交通中同样重要。

标识系统

根据指引需要的不同，公共标识系统的差别也很大。一个标识系统可能是为快速通过某个地点的大量人流服务，也可能是另一个极端，只为某一个人在某个环境中找到某个位置提供帮助。与帮助人们找到某个地点同样重要的事情是帮助

ORIENTATION

One prerequisite for a functional and pleasant pedestrian environment is good orientation in the area. It is important for the pedestrian to know exactly where he is at any moment and the direction to the most important destinations. Two main types of orientation can be distinguished: 1. the spacial shape of the environment and the awareness of location and direction created by what is visible and 2. orientation created by signs.

For direct orientation the following factors are important:

– Clarity of route. A grid-plan network is good in this respect.

– Diagonals weaken orientation. It is difficult for people to assess changes in direction that are other than right angles.

– Lack of symmetry help to express direction.

– Symmetry and articulation make for easier assessment of the dimensions of spaces.

– Orientation is improved by application of the central space principle, i.e. using large areas from which one can get a general picture of the surroundings (e.g. squares, central halls, etc.)

– Landmarks, buildings, topographical forms, etc. visible for a long way from several directions help one to assess direction and distances.

– Lighting. Points important for orientations should be emphasised. Traffic channels are in a key position: junctions, exchanges, terminals are particularly important (clear separation of traffic flows important).

When the street network is well-arranged and fairly regular in form it is fairly easy to move around and to sense one's location in relation to the whole. Homogeneity and lack of street hierarchy in an area lead to difficulty with orientation, and then a sign system is vital.

GUIDANCE

As public transport continues to grow, more attention is being given to signs aimed at providing information to the public about transport services.

Ergonomic features have come to the forefront in the design of these signs. It has been found essential to put the supply of information into the best form for people to grasp and one which cannot be interpreted ambiguously. In public transport both visual and auditive information is given.

SIGN SYSTEM

The demands made of a public sign system vary greatly, depending on the situation in which guidance is needed.

A sign system may serve huge numbers passing rapidly by some point or, at the other extreme, just one person trying to find a given place in some

人们怎样离开某处去别的地方。无论怎样的情况，所有的标识应该以一种统一的基本模式为基础。

如果我们考察作为环境元素的标识系统，以及标识与人的感知能力的关系，就会得出以下一些原则：

1、信息的内容：

——信息应容易理解

——信息应正确

——信息不能模棱两可

2、信息的数量和位置：

——信息量应充分

——信息应及时

——信息的位置应正确

3、信息的重要性：

——在不同地方提供信息的标识应该尽量类似（提供信息的方式应该相同）

——必须合理有效地提供信息（比如，运用某个容易看懂的符号或大号的字体）

设计标识系统

标识系统的设计应该与其它部分的设计同步进行。当建筑或建筑的一部分开放使用时，标识系统也应该到位。一个好的标识系统需要经过精心设计。即使在运用标准符号系统的情况下，它们的具体位置和内容也应该经过规划。复杂而未经设计的信息传递将造成标识系统的混乱。

如果建筑师在规划和建筑设计中能够考虑到标识的问题，将有助于建立一个良好的识别系统。但常见的情况是，他并不具备这方面的知识而需要求助于艺术设计师。（在芬兰，艺术设计师本人在这方面的训练往往也不够。）

设计师同样也应考虑材质的运用。选材需要考虑的因素有：材料是否容易清洁、系统寿命是否长久等等。

标识系统的视觉协调

标识和信息系统的视觉协调指的是系统应该使用相同的基本组件。例如：时间表和其它的标识牌应该使用相同的字体，以及设计应该运用统一的图形规则等等。这些原则适用于字体、色彩和符号等基本视觉元素。元素的选择和设计需要受到特别的关注，因为它们运用广泛，直接关系到信息系统的清晰程度。

organization. Just as important as guidance to some point is guidance away from it or to some other part of a building or another building. However, all signs should be on one basic pattern within a homogeneous system.

If we examine the sign as an element of the environment and the sign's relations with the person seeing it, we can lay down the following demands:

1. Content of information

– the information should be easy to understand

– the information should be correct

– the information should be unambiguous

2. Amount and location of information

– the information should be adequate

– the information should be correctly timed

– the information should be located in the proper place

3. The significance of the information

– the signs giving the information in various places should be similar (the way it is given should be the same)

– the message should be correctly given in the functional sense (e.g. a symbol that can easily be interpreted or sufficiently large lettering)

PLANNING A SIGN SYSTEM

The sign system should be designed at the same time as other planning. When a building or part of it is opened, the signs should be in position. A good sign system calls for careful planning. Even if standard signs are used, their location and content should be planned. Jungles of signs are often the result of complicated unplanned communication.

The best sign systems are produced if the architect planning the area or building can also study the sign system. Often, however, he is not adequately trained for this purpose and has to use an artist (who may not be adequately trained for the job either, here in Finland).

Planning should also pay attention to the materials used. Factors affecting choice of material include how well it can be cleaned and the system's lifespan.

VISUAL COORDINATION OF SIGN SYSTEM

Visual coordination of the sign and information system means that the system should use the same basic components, e.g. timetables should use the same kind of lettering as the signs, or the graphic rules applied to their design should be the same. This applies to all basic visual elements such as lettering, colours and symbols. This presupposes careful choise and planning of the elements concerned, because they are used often and in many places and affect the clarity of the message and the public image of the whole system.

小中有美

艾里克·布里格曼的别墅建筑

在芬兰这样的小国家，大型建筑常常受到尊崇，而对所谓大型建筑的界定又常常只凭尺度和数据。并没有绝对的标准。小的建筑可能与大的建筑一样重要。人本身才是衡量一切的唯一尺度，而尺度又是美的主要元素之一。

私人住宅在建筑师的作品中有着特殊的地位。因为在设计私人住宅时，建筑师可以不受外界压力和规定的影响，不需要过多的妥协，建筑师得以在最小的风险基础上试验新的风格和新的结构细部。在私人住宅里，房间与空间的设计更易把握，因为理想住宅存在于我们每一个人自己心中。在小住宅设计中，建筑就是人类现象，它告诉我们人是什么，而建筑师也最易通过它来表达自己的个性。

艾里克·布里格曼设计的别墅是他作品中的关键部分，同时也是芬兰建筑风格与潮流的生动表现。在本世纪的上半叶，芬兰的潮流与世界潮流同步。在所有的私人住宅设计项目里，布里格曼忠实于自己的信念，而只有从自己的信念出发的设计才能上升到建筑的水平。

本文的主题是通过布里格曼的住宅设计来探讨他的建筑。住宅是建筑师的试验场，它使人们从很小的尺度上检验现代设计的潮流。在这些项目里同样可以看到建筑技术革新和改造。不过，从这方面看，维特鲁威建筑的原则之一的"坚固"不幸需要让位于其它的因素。

小住宅通常不会给它们所处的环境带来剧烈的变化。正因为如此，建筑师可以在他的作品中作些试验，尽管住在里面的人可能会因为这些试验而忍受生活中的某些不便。布里格曼也许称不上是一位先锋派或改革派的建筑师，但在他的职业生涯里，他一直都热衷于尝试新的东西。他处理设计问题的个人风格清楚地表现出他对新的可能性的兴趣。在这个意义上，布里格曼与阿尔瓦·阿尔托一样走了一条功能主义建筑的道路。

和巩拿·阿斯普朗德的作品一样，布里格曼的设计来自直觉而非理论与空想，因此两位建筑师设计的别墅有许多共同的地方。当功能主义在芬兰建筑界获得突破性地位的时候，布里格曼这样描述建筑师的职业："这不是一种简单的数字工作，由于其中众多的不确定因素，它需要发挥创造性直觉找出最佳方案。"

布里格曼的目标是为那些与自然保持着亲密联系的人们创造出美好的环

Small is beautiful [1]

Erik Bryggman's villa architecture

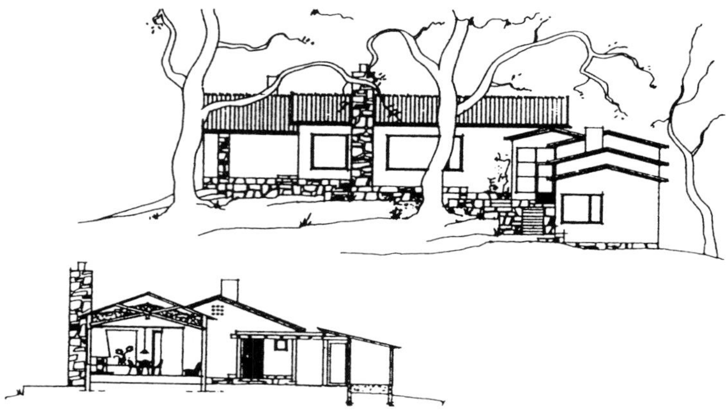

In a small country such as Finland, largescale building projects are often admired and valued, and their status is often based on the sheer statistics. There are, however, no absolute criteria of size. The very small can be just as important as the very big. Man himself is his only measure, and scale is one of the main elements of beauty.

Private houses usually have a special role among the works of architects. They represent tasks exceptionally free of outside pressures, regulations or compromises, permitting the architect to experiment with new styles and structural details with a minimum of risk. The design of rooms and space is easy to command in a private house, for the ideal of a house is within all of us. In small houses, architecture is a human phenomenon, telling us of what it is to be human, and in such projects the architect most clearly expresses his own personality.

Erik Bryggman's villas are a key to his creative work, and also a living story of the trends of style in Finnish architecture. in the first half of this century, these trends have actively followed international currents. In all of his projects for private houses, Erik Bryggman remained faithful to his own views, for only a personal approach to a project will raise it to the level of architecture.

The subject of this article is the development of Bryggman's architecture as demonstrated by his designs for private houses. The villas were a testing ground for the architect, permitting a study of contemporary trends on a small scale. Innovations and reforms in building technology were also evaluated in these projects. In this respect, however, the Vitruvian principle of "*firmitas*" has unfortunately had to give way to other considerations.

Small houses do not usually involve any drastic changes to their environment. In this respect, the experiments were fortunate for the architect's later works, although some times those who live in the houses have had to suffer some inconvenience because of them. Erik Bryggman can hardly be regarded as a forerunner or reformer of architecture, but he was actively interested in new developments throughout his career. His personal approach to all of his projects and commissions clearly showed his interest in new possibilities. In this sense, Bryggman and Alvar Aalto followed the same route into functionalist architecture.

Like Gunnar Asplund's works, Bryggman's architecture was based on artistic inspiration and intuition rather than theory or speculation, and there are many common features in the villas designed by both architects. When functionalism was making its breakthrough in Finnish architecture, Bryggman defined the task of the architect as follows:

"*It is not one of simple calculation, but, with its many unknown factors, it requires creative intuition to be solved in the best possible way.*" [2]

Bryggman's aim was to create an environment suitable for individuals that remained in touch with its natural setting. He was clearly adverse to monumentality, pathos and large-scale planning. Erik Bryggman's works involve intimate proportions, free composition and complexity, all of which

境。他明确地反对纪念性的、大尺度的工程。布里格曼的建筑有着精妙的比例关系、自由的构图，与含蓄的多样性，这些品质产生出丰富而生动的效果。他的建筑的尺度、比例和形式关系表现出感性而复杂的一面。

建造住宅的原则与建造巢穴的原则相类似。设计私人住宅时，建筑师必须将自己放在居住者的位置上，并融入自己的个性。小住宅的设计同样反映出建筑师与自然和环境之间的关系。

根据传统建筑的设计原则，最适宜布置住宅的地点也是小气候环境中最有利的位置，同时又拥有最好的视野。即使在布里格曼设计中的功能主义时期，他一直遵从着这种传统的布局原则。

布里格曼设计的平面并不十分与众不同，它们主要是以传统的和多少年来已经广为人们接受的基本做法为依据。他的几乎所有的别墅设计都重复着同样的布置方式：将主要房间（起居室、餐厅和厨房）与辅助房间或空间（卧室等）的位置相互垂直布置。而当他将这种基础的平面形式与自然环境相结合时，就表现出他出众的想象力。这方面最好的例子是努提拉别墅，在这个设计中环境与总平面布局产生了决定性的效果。

布里格曼能够很快地在立面设计上采用新的东西。他早年的别墅表现的是民族浪漫主义的风格，但是，当他从意大利之行和泰瑟诺的书籍里获得新的印象之后，布里格曼的设计开始表现古典主义原则（斯哥斯普和索林别墅），其中也包含了某些早期的功能主义痕迹。而在30年代的别墅和竞赛设计中（华伦别墅、艾克曼别墅、凯诺别墅和贾拉农别墅），可以看到纯净的功能主义的充分表现。

布里格曼后期的别墅建筑转向了浪漫主义。努提拉别墅综合了他的许多特点。在这项工程里，成熟的建筑经验与感性的艺术直觉相融合。事实上别墅是评价布里格曼一生的建筑作品的关键依据。通过它们人们可以清晰而全面地描述和理解布里格曼在创作中所体现出来的艺术气质。同时，在这方面的研究也反映出近现代芬兰建筑与建筑风格的发展情况，这段时期的芬兰建筑深受国外影响。

1913年，在建筑学院学习的第三年，布里格曼参与了一项奥塔瓦出版公司组织的避暑别墅的设计竞赛。当时避暑别墅或乡村别墅这一类型的建筑还

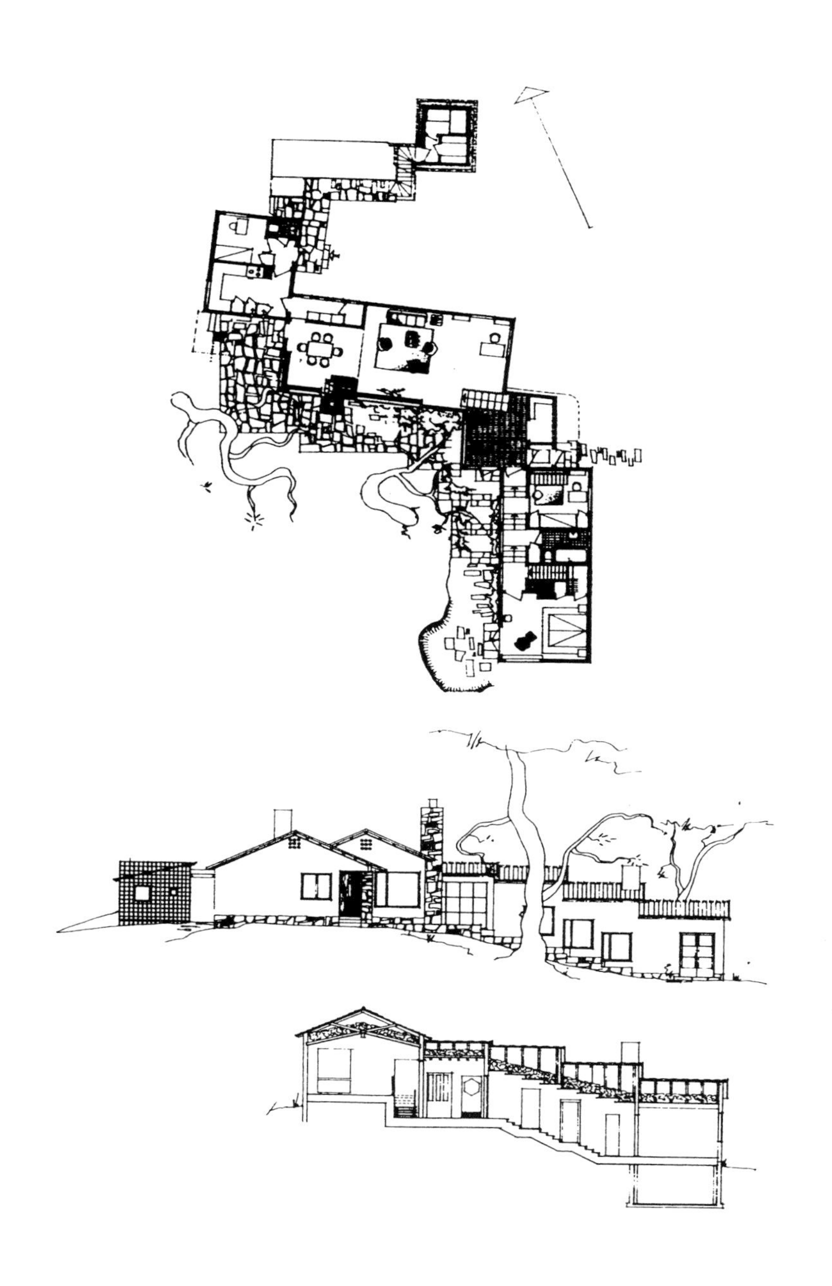

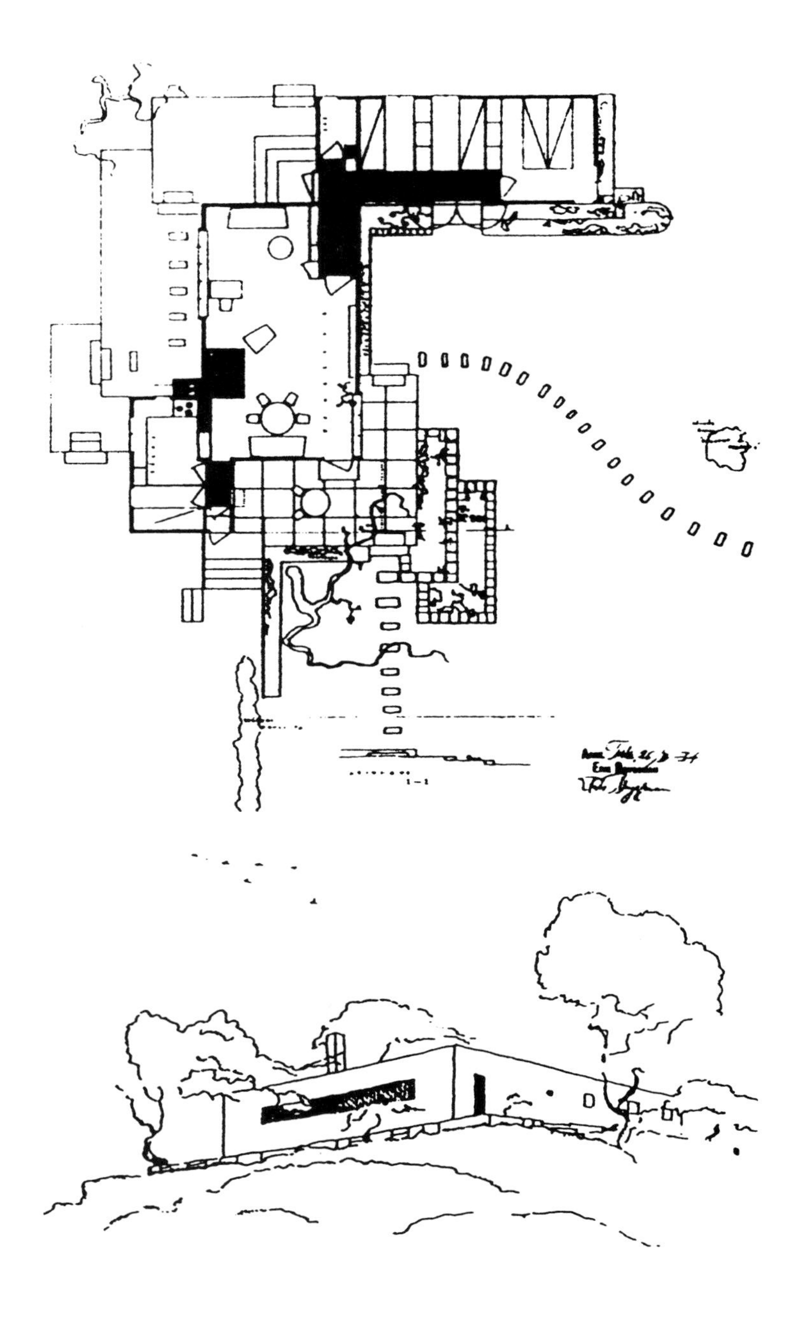

led to rich and vivid results. Along with scale, proportions and forms were sensitively comprehended features of his architectrue.[3]

An apartment involves the same principles as building a nest. In private houses, the architect has to place himself in the role of the inhabitants and involve his own personality. The design of small housing also reflects the architect's relationship with nature and the environment.

According to the old tradition of vernacular architecture, the best place for the house was the most suitable one in terms of microclimate, but it may also have been in the ugliest part of the lot, providing the most beautiful view of the surroundings. Bryggman followed this tradition in the siting of his villlas, even during his functionalist period.

Bryggman's villas were not especially radical in plan, but were based on tradition and a basic articulation that had been learned and had come to be accepted over the years. Almost all of his villa designs repeat a basic configuration of main rooms (living room, dining room and kitchen) with auxiliary rooms and space (bedrooms etc.) located at right angles to them. He was equally imaginative, however, in fitting this basic plan to the natural surroundings of his projects. The best example of this is Villa Nuuttila, where considerations of landscape and siting had a decisive effect on the result.

Bryggman was quick to adopt new influences in his façade designs. His earliest villas were still in the national-romantic style, but new impressions from his Italian journey and the books of Tessenow[4] led to designs based on the principles of classicism (Villa Skogsböle and Villa Solin), in which incipient functionalism also played a part. His purest expressions of functionalism can be seen in the villas and competition entries of the 1930s (Villa Warén, Villa Ekman, Villa Kaino and Villa Jaatinen).

Bryggman's later villa architecture branched off towards romanticism. Villa Nuuttila represents a synthesis of his architecture. In this project, mature architectural experience is combined with sensitive artistic intuition. The villas are, in fact, key works in assessing the whole of Bryggman's architectural output. They permit an overall review of his creative work in a way that may describe and explain his artistic temperament clearly and unequivocally. Such a review also offers a brief survey of the recent history of Finnish architecture and its styles, which were mostly derived from abroad.

In 1913, while in his third year of architectural studies, Bryggman participated in a competition for the design of a summer villa, arranged by the Otava publishing company.[5] At this time, there were no independent forms or conventions for summer villas or country houses, and they were mainly based on features from urban villas and other types of small housing. Bryggman's entry, in the national-romantic style of the day, was awarded a shared third prize. The project was the subject of much interest, and a perspective illustration of the entry was published on the cover the *Otava* magazine's number devoted to the competition.[6]

Finnish architects of the 1920s were greatly interested in vernacular

没有形成自己独立形式或通用的做法，这类别墅的设计主要是从城市独立住宅和其它小型的住宅类型中演变而来。布里格曼的设计在民族浪漫主义风行的当时，获得了并列三等奖。布里格曼的设计方案引发了人们广泛的兴趣。它的透视图被刊登在以那一次设计竞赛为专题的奥塔瓦杂志的封面上。

1920年代，芬兰建筑师对乡土建筑产生浓厚的兴趣。民族浪漫主义思潮的盛行使人们更尊崇传统建筑粗犷、简洁的风格。而渐渐地，早期的牧屋式风格又被强调功能和实用的新客观主义风格所取代。

在这样的背景下，1924年，布里格曼设计了代表乡土和传统主义风格的斯哥斯博农场住宅。

场地上原有的两个建筑物决定了新建筑的性质和范围。布里格曼在这里首次运用了横向的设计，形成了一个避风的庭院。其设计灵感显然来自于阿斯普朗德于几年前设计并已建成的斯耐尔曼别墅。建筑狭长的形体创造出宽敞轻盈的空间感。这一长向的体量容纳了贯穿在整个建筑当中的大厅和起居室，同时其它的房间又有很好的景观。主入口的屋顶处理采用了细腻的传统形式。其细部，包括别具一格的弯曲的柱子，在布里格曼以后的作品中不断地出现。

1929年，根据布里格曼的设计，在靠近图尔库的卡特林南拉斯科建造了索林别墅。设计开始于1927年，早期的设计包含了许多传统主义的主题和装饰。在修改的过程中，设计逐渐趋向简单，表现出功能主义理论的影响。这一作品多方位地表现出从泰瑟诺、赖特、阿斯普朗德的设计中获得的影响，同时，又是斯哥斯博农场住宅中许多设计原则的直接延续，并融入了新的做法。

在索林别墅中，布里格曼再次使用了狭长的横向体量，其中建筑的侧翼是一个有顶的室外空间，三面受到围合。在敞廊这部分的设计中，布里格曼尝试了功能主义的形式元素。原来的坡屋面改成平屋面，这样就为第二层提供了一个屋顶平台。实际上，索林别墅中包含了非常丰富的设计思想。

建筑位于两座山之间的皮卡索米海峡岸边，这是一个相当传统的别墅地形。最佳视野是从屋顶平台和敞廊面向花园和大海。住宅的另一边是一个庭院，庭院由一个与主体建筑平行的独立小屋，和靠近路边的矮墙围合而成。独立小屋被一个门形的构筑物分为两部分，这是布里格曼在多个设计中都使用

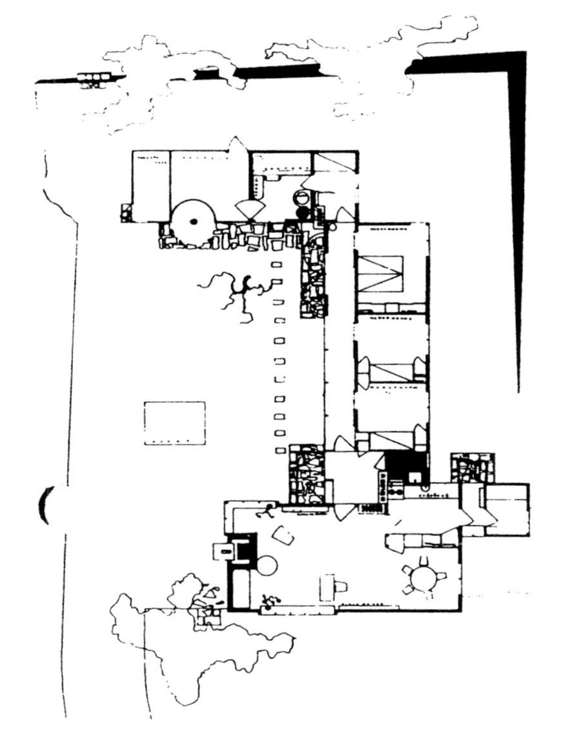

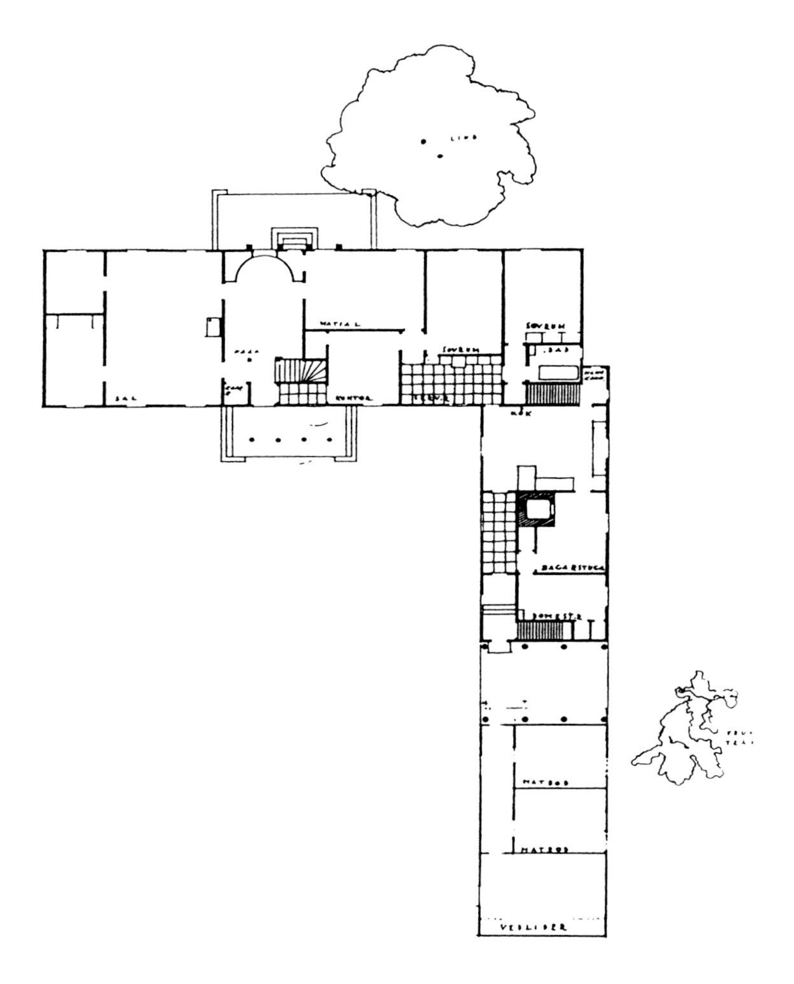

architecture. The idealism of the national-romantic style led to an appreciation of the rough and unadorned simplicity of traditional styles of building. The earlier chalet style gradually gave way to *Neue Sachlichkeit* and a stress on the practical and functional features of building.

It was within this context that Bryggman designed in 1924 the Skogsböle farmhouse in Kemiö,[7] an example of vernacular traditionalism among the architect's works.

Two buildings already standing at the site dictated the nature and limits of the project. This was the first time that Bryggman employed a transverse design, creating a yard shielded from the winds. A clear source of inspiration was Gunnar Asplund's Villa Snellman, which had been built only a few years earlier.[8] The narrow body of the building was a feature creating a feeling of spaciousness and lightness. It also permitted a hall and a living room passing through the body, and the rooms of the interior offer beautiful views of the surroundings. The roofing over the main entrance is a refined version of porches in vernacular architecture. Its details, including the markedly curved forms of the pillars, were to be repeated in many of Bryggman's later works.

In 1929,Villa Solin was bulit at Katariinanlaakso near Turku according to Bryggmans's designs.[9] The planning of this villa had begun in 1927, and the early stages still contained many themes and ornaments derived from classicism. The design became simplified and gradually began to display the influence of functionalist theories. Upon completion, the villa was a synthesis of prototypes derived from Tessenow, Frank Lloyd Wright and Gunnar Asplund. It was also a direct continuation of the principles applied in the Skogsböle project, with an infusion of new trends.

In Villa Solin Bryggman again used a narrow-bodied transverse design, in which the wing of the building, a roofed outdoor space, is protected from three sides. Bryggman experimented with functionalist forms in this loggia part. The original hipped roof of the wing was replaced by a flat roof, which could also be used as the terrace of the upper storey. Villa Solin, in fact, combines a variety of architectural ideas in a single project.

The location of the building, between two hills on the shore of Pitkäsalmi Sound, was a traditional one. The best views were from the terrace and the loggia towards the gardens and the sea. On the opposite side of the house was a yard, flanked by an outbuilding located parallel to the main builing and a low wall adjoining the nearby road. The outbuilding is divided into two parts by a gate structure-a theme used by Bryggman in a number of designs. Villa Solin has fallen into disrepair over the years, but the City of Turku is now planning renovations, and the villa will be used to accommodate official guests of the city. We hope that the building will be restored to its original appearance.

In 1932, two architectural competitions were held in Finland for the design of smallscale housing.[10] The Enso-Gutzeit firm arranged a competition for the design of a weekend cabin. This was partly to investigate the possibilities of

的一个题材。索林别墅曾经年久失修，图尔库市政府目前正在计划修复。修复之后，索林别墅将会被用作城市官方的招待所。我们希望建筑最终能恢复原样。

1932年，芬兰进行过两次小型住宅的设计竞赛。其中恩索－古采特公司组织的是度假别墅的设计。这次竞赛的一部分目的是为了研究一种新型纤维板——恩索板在这类建筑中的运用，同时寻求发展出一种结构简单但质量优良的建筑类型。组委会将会从入选方案中挑选一个，在1932年北欧建筑会议的露天展场建造。竞赛包括两个系列，A系列为25m²至35m²面积的临时性小住宅。B系列为较大的建筑（50m²至60m²），拟作较长期的使用。

所有的入选和获奖方案基本都遵从了功能主义的设计原则。两次竞赛都由建筑师希托能和路柯能合作设计的较为一般的作品获胜。从现在看来，艾里克·布里格曼和阿尔瓦·阿尔托的设计要有趣得多。他们俩都因系列A的设计而获奖。除此之外，布里格曼的系列B的设计因违反竞赛要求而未获参赛资格，但在竞赛之外设计被人购买。布里格曼的系列A的作品标题为"多罗茜·柏金丝"。评委们对他的作品的评价是："整体上这是一个感性而优美的设计，但造价较贵，尺度过小，建筑中包含两个烟囱和三个对外的门。房间的布置清晰，值得赞赏。立面设计优雅但过于复杂。"同一年，芬兰的茵苏莱公司为北欧建筑会议组织了另一个名为茵苏莱别墅设计的竞赛。布里格曼获四等奖。这次竞赛的胜者为建筑师欧·弗罗丁。与其它几次竞赛一样，这次的获胜设计同样是有功能主义的特征。

布里格曼的设计（题为"X"）获得了如下的评语："厨房没有独立的出入口，起居室显得较为嘈杂。起居室和卧室的入口占了大约4个平方米，这部分面积本应该用于门厅。厨房的设计没有完成，并且缺少杯具台。立面优美。屋面倾角太小，不利布置屋顶结构。"

这些竞赛作品清楚地表现了功能主义的特征，为布里格曼30年代的设计打下了基础。同时，这些作品体现了设计者的个性和习惯，它们本身也构成了一个延续、完整的系列。华伦别墅是这些作品中最独特、最引人注目的一个，和努提拉别墅一样，华伦别墅是布里格曼最著名的作品之一。

华伦别墅的布局考虑到景观和海洋主导风向的因素。布里格曼在此运用

using a new fibreboard material, Ensonite, in buildings of this kind. The competition also aimed at developing type housing that was strusturally simple and of high architectural quality. One of the entries was to be erected at the outdoor exhibition of the Nordic Building Conference held in 1932. The competition involved two series. Series A was for cabins of 25–35 square metres in floor area that were intended for short-term use. Series B was for slightly larger buildings (50–60 square metres), designed for longer occupation.

All of the prize-winning or remunerated entries were more or less functionalist in spirit. First prize in both series went to the architects Hytönen and Luukkonen with their relatively standard designs. In retrospect, the entries by Erik Bryggman and Alvar Aalto were much more interesting, and they were remunerated in series A. In addition, Bryggman's entry in series B was purchased from outside the competition, as it was contrary to the expressed programme and had to be disqualified. Bryggman's entry in series A under the pseudonym "*Dorothy Perkins*" was described by the jury as follows: "A sensitive and beautiful overall approach. The entry is, however, costly and too small in scale. – Two chimneys, three outside doors. Configuration of rooms is clear and praiseworthy. – Beautiful exterior, but too complex." In the same year, the Insulite Company of Finland arranged the so-called Insulite Villa Competition, also for the Nordic Buliding Conference. Bryggman received fourth prize. The winner was the architect O. Flodin. As in the other competition, here too the prize-winning entries were of a functionalist character.

Bryggman's entry ("X") was commented upon in the following terms: "There is no separate entrance to the kitchen. The living room appears restless. The entrance to the living room and the bedroom occupies some four square metres of floor space, which could have been used in the entrance hall. The design of the kitchen is incomplete, and there are no cupboards. Beautiful façades. Roof angle too small and insufficient space for roof structrues."

These entries were the predecessors of Bryggman's villas of the 1930s, clearly expressing the features of functionalism. However, these works also reveal the architect's personality and his tradition to such a degree that they can be seen as a uniform series. Villa Warén[11] is perhaps the most unequivocal example, and in a way the most intriguing one. Along with Villa Nuuttila, it is perhaps the best publicized of Bryggman's villas.

The location of Villa Warén takes into account both the landscape and the prevailing winds from the sea. Here too, Bryggman used his familiar plan. Placed at a right angle to the living room and the dining room is the bedroom wing, forming an extension of the former part. Added to the yard is a service wing, which has later been used as a dwelling. The semicircular recess of the service wing, sparing a pine growing at the site, shows a clear concession to functionalism, but also an approach typical of Bryggman.

了他所习惯的方法，将起居室和餐厅区域与卧室区域成垂直关系布置，构成对原有建筑的延续。庭院附近增加了一个服务性建筑体，后来也被用作居住。这个服务性的建筑在形体上作了一个半圆形的后退，避开了地形位置上的一棵松树，这一处理体现出清晰的功能主义特点，同时也是布里格曼典型的手法之一。

华伦别墅与其它30年代的别墅一样，主体框架为木结构，外部为灰泥涂料，涂料在许多年后出现了一些问题，业主也因阁楼和地下室的通风不够而备受困扰。与那个时期的其它一些住宅的做法一样，布里格曼为华伦别墅的起居室设计了一个高于建筑其它部位的坡屋顶。

位于赫文沙洛和卡斯克塔的埃克曼和凯诺别墅是华伦别墅的变体。埃克曼别墅设计的一个有趣之处是有一堵混凝土墙将花园和庭院的其它部分分离开来。在埃索周末别墅竞赛的获奖设计中，布里格曼也曾用过类似的手法。

1939年在普佛群岛建成的贾提能别墅是布里格曼30年代设计中最大的一个。它位于陡峭的悬崖边，正面临海，背面朝北，是一片森林。别墅按通常的做法分成两部分：较大的部分容纳了起居、餐厅和厨房，较狭窄的一部分为卧室。稍有不同的是，贾提能别墅中两者在形体上相互分开，通过一条有顶的通廊相联系。敞廊又成为有顶的户外活动区。从柱与敞廊的基础所用的未经加工的石料，以及其它的一些手法中可以看出，在贾提能别墅里，严谨的功能主义逐渐让位于一种更为含蓄而浪漫的建筑风格。

1949年在库斯托建成的努提拉别墅也许是布里格曼最著名的别墅。这项工程原来由埃里克·豪森委托，后因资金问题而放弃。安提·努提拉后来买下了这个未完成的项目，但不能对它作任何更改。别墅的基地位置有一块巨石，必须移开。起居室位置的前方有一棵巨大的橡树，因此平面的网格有所调整，使起居室面向最佳的景观朝向：皮卡索米海峡。古橡树和巨石使得这个地点具有别具一格的风貌。建筑原来为避暑别墅，但也可以常年使用。

入口的处理尤其引人注目。门厅处展现出周围的景观，同时可以看到几级踏步之外的起居室。处于较低位置的卧室也可以通过门厅到达。起居与餐厅形成统一的大空间，这一大空间的中心是外墙中央位置的一个壁炉。餐厅的后部是厨房及厨房服务区，并面向庭院开门。如果风向条件有利，厨房后

Villa Warén, like the other villas of the 1930s, had a wooden framework with a stuccoed exterior. The stuccoing has caused a number of problems in later years. The owners have also been troubled by insufficient room for ventilation in the attic and the basement. In Villa Warén, as in other villas of this period, Bryggman designed an oblique ceiling for the living room that was higher than in the rest of the house.

Villa Ekman[12] and Villa Kaino[13], in Hirvensalo and Kakskerta, are variations of Villa Warén. Of interest in Villa Ekman is the wall of concrete separating the garden kitchen from the rest of the yard. Bryggman had used similar features in his remunerated entry for the Enso-Gutzeit weekend cabin.

Villa Jaatinen,[14] built in 1939 in the Porvoo archipelago, is the largest of Bryggman's villas of the 1930s. It is located at a site adjacent to steep cliffs facing the sea, with a background of forest to the north. The villa is typically divided into two sections: a larger part containing the living room, a dining area and the kitchen, and a narrower part with bedrooms. In Villa Jaatinen, however, these parts are kept separate, and are joined by a roofed walk. The loggia also forms a roofed outdoor area. In Villa Jaatinen , the severity of functionalism gave way to a more subdued and romantic style of architecture, as shown, for example, by the use of undressed stone in the pillars and foundations of the loggia.

Villa Nuuttila,[15] which was built in 1949 in Kuusisto, is perhaps Bryggman's bestknown villa design. It was originally commissioned by Erik Hausen, who had to abandon the project for financial reasons. Antti Nuuttila purchased the unfinished building, and thus was not able to have any say in the result. The villa was placed at the site of a large boulder, which had to be removed. A large oak tree in front of the living room window led to changes in the grid of the floor plan, and the living room opens towards Pitkäsalmi Sound and the best views available from the site. The large oak trees and boulders give the location a dramatic appearance. The building was originally intended as a summer villa, but it can be used throughout the year.

The entrance is especially impressive. the hall provides a view of the surrounding scenery and the living room ,which is reached by a few steps. The bedroom wing, at a lower elevation, is also reached from the hall. The living room and dining area from a single space, dominated by a fireplace in the middle part of the outer wall. At the rear of the dining area is the kitchen with its serving area and access to the yard. Depending on the winds, the yard behind the kitchen can also be used for dining outdoors.

The picture window in the living room shows the gradated wall of the bedroom wing, the large oak tree and a boulder, with its surface texture changing according to the seasons. The master bedroom is well lit, and there is a bathroom between this room and a smaller bedroom. The floors of the entrance hall, the corridor of the bedroom wing and the stairs are laid with tiles, while the floors of the living room and dining area are of boards. This late villa design by Bryggman is a mature architectonic creation, where all of

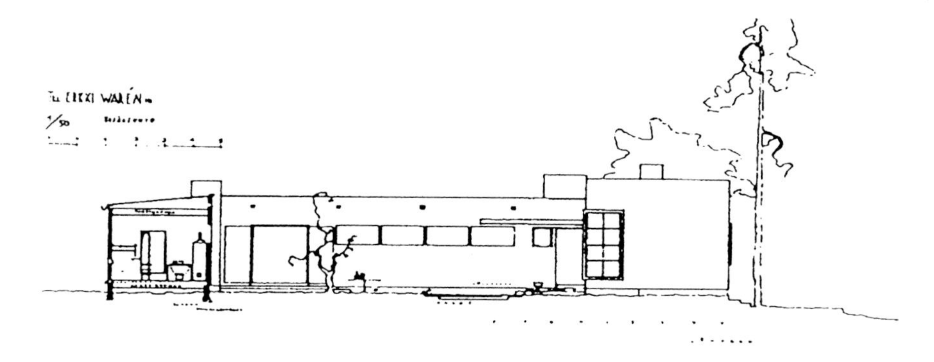

的庭院也可用作室外的餐厅。

从起居室落地窗望去，可以看到卧室渐变的外墙，那棵橡树和巨石，巨石的表面肌理随着季节的变化而变化。主卧室光线充足，在主卧室与另一卧室之间设有一个洗手间。入口门厅，卧室走道和楼梯位置铺地砖，起居室和餐厅为木地板。布里格曼后期设计的别墅是成熟的建构创作，维特鲁威的要素在此恰如其分地得到体现。虽然立面用的是涂料，但由于出挑屋檐的保护而未受到过多侵蚀。建筑主体为木构，以一种叫伯格纳板（一种高压纤维板）的材料做面层及保温隔热，板外为涂料层。

屋顶为瓦屋面，原来的双沟屋面瓦现在已经不再生产，给维修带来问题。室内设计由布里格曼的女儿卡琳·布里格曼协助完成。1951 年，根据布里格曼的设计在靠住宅边的位置增加了一个桑拿室。桑拿室的外立面反映出布里格曼早期对意大利民间建筑的强烈兴趣。从整体布局到室内细部处理，这座别墅充分表现了布里格曼的设计特点，是他成熟期最重要的作品之一。希望这一作品能够永远保存下来，让后代了解当时优秀的设计和高质量建造成果。

别墅设计形成了一个小型的建构世界，它使人们了解到建筑师作品中引人入胜的方面。今天，当我们面对媒体给我们带来的形形色色的建筑取向和观点的时候，埃里克·布里格曼的别墅建筑更加永恒而令人信服。简洁、人体尺度和环境这些因素是现在的建筑师同样需要认真考虑的问题，从环保和生态角度对建造手段进行思考不仅是一种可能性更是一种迫切要求。在混杂着后现代、晚期现代和其它种种现代主义风格的当今建筑界，布里格曼的别墅所表现出来的优秀品质清新而隽永。现在，遵从自然的设计时代又一次到来，尽管这个时代深受疯狂的消费主义所侵扰，但人们已经作好准备去追寻一条更为清醒的建筑道路。

小可能重新受到赞美。

（注释略）

the Vitruvian factors appear to be in their right places. Although the façades are stuccoed, they have been well preserved because of the overhanging eaves. The building has a timber framework, covered and insulated with so-called Bergner board (a rigid boarding material of compressed reeds), upon which the stucco was applied.

The roof is of tiles. The original twin-furrowed roof tiles are no longer made, and repairs to the roof have led to problems. Bryggman was assisted in the interior design by his daughter Carin Bryggman. In 1951, a sauna was built at the shore according to Bryggman's designs. The exterior of the sauna is strongly reminiscent of Bryggman's early interest in the "*architettura minore*" of Italy. From its location to the details of the interior, the villa is a total work of architecture and one of the main works of Bryggman's mature period. It is to be hoped that it will be preserved for future generations as an example of high-quality design and construction.

The villa, as an architectonic world in miniature, provides an interesting perspective on the works of the architect. Today, when different architectrual orientations and views are constantly being thrust upon us by the media, the features of Erik Bryggman's villa architecture seem permanent and reassuring. Simplicity, a human scale and respect for the natural setting are all criteria that even today's architects should take into account. In the future, an ecological attitude to construction will be more of an imperative than just a possibility. Amidst the present flurry of post-, late- and other modernisms, the best qualities of Bryggman's villas seem refreshing and timeless. The time of architectrue attuned to nature is again at hand, and our modern age, plagued by hysterical consumerism, may yet be ready for a cooler approach to architecture.

Small may again be beautiful.

NOTES

1. Schumacher 1973.
2. Granskaren 1932, 93.
3. Sarjala 1964.
4. Tessenow 1916.
5. ARK 1913, 49.
6. Stigell 1965, 11.
7. ARK 1929, 72-74.
8. Asplund 1943, 89.
9. ARK 1929, 91-92.
10. ARK 1932, 63, 95, 97.
11. ARK 1933, 156-157.
12. ARK 1933, 157-158.
13. ARK 1935, 172
14. ARK 1953, 77-79.
15. ARK 1952, 151-157.

交通建筑中的钢与玻璃

Jack Kerouac 一定知道公路上的情形。这位后现代主义者没有必要知道为什么他一直在旅行。旅游在今天已成为一个最大的产业，从而交通站点就成为我们这个时代常常光顾的地方。

在1993年世界状况报告中，有一篇题为“铁路运输的回归”的文章，它描述了很多欧洲国家正在计划在不远的将来提高铁路运输能力。法国率先在20世纪80年代就开始行动了。德国计划在2010年以前相比较公路而言，在铁路方面增加更多的投资。瑞典则计划在以后十年中，在铁路上的投资至少与公路上的投资相持平。

铁路运输可以节约能源，同时它还比公路运输安全。它也常常比我们过去想像的更加方便快捷。

在发展公共运输方面的一个重要理念就是平滑。因此我们应当设法消除阻碍交通建筑设施平滑的因素，而增加建筑的透明度。终点站的建筑变得更加生动有活力，但它们必须保证给乘客提供足够的私有空间。外面行人的影响对我们来说显而易见。

旅行的吸引力之一在于它的同时在移动与在目的地的思想。它正好是透明理念的一部分。在这儿的同时又在那儿。透明性实现了这样两个方面。钢与玻璃一起创造了这种可能，实现了这种幻觉。

钢与玻璃所实现的轻巧正是这个千年遇到的挑战之一。Buckminster Fuller 一直强调建筑的重量。根据新的生态理念，这也包含了在构造加工过程中消耗的能量，以及维护这座建筑所消耗的能量。

在交通建筑设施中，第三个千年的巨大挑战可能是轻巧、生动、透明、采光、速度、精确以及可靠。

然而在旅行中，最重要的事情就是要能非常容易地找到下一个目的地。那就是为什么标记是交通建筑的一部分的原因。空间和环境本身对你的引导越多越好。标记能提高空间可视化的质量。等待是旅行的一部分，然而等待的舒适性没有得到充分的考虑。当考虑等待时，休息室的设计原则变得非常重要。

在实际的运输工具中，两个基本点必须加以强调。交通工具必须要体现生动有活力，同时还要提供舒适。这意味着运输工具的设计面临着更加严峻的挑战。

在19世纪末期，铁路的蓬勃发展可以从火车站的迅速增加得到体现。在

Steel and Glass in Traffic Architecture

Jack Kerouac must have known what it is like to be on the road.

The post-modern person does not necessarily always know why he is travelling. Tourism is the largest industry today and traffic terminals are the cathedrals of our time.

The State of the World 1993 report contains an article "The Return of Rail Transit", which states that several European Governments are planning improvements on rail traffic in the near future. France acted as a pioneer as early as the 1980's. Germany is expected to invest more on railways than roads by the year 2010. Sweden plans to invest at least as much on railways as on roads in this decade.

Rail transit saves energy. It is also safer than road traffic. It could also be considerably more convenient than what we are used to.

An important concept in the development of public transport is smoothness. This is reflected in removing obstacles that impair the smoothness of traffic architecture and by increasing the transparency of the architecture. The architecture applied to the terminals is becoming ever more dynamic, but they also need to supply a sufficient degree of privacy to the passengers. The role of an outsider also comes naturally to us.

The attraction of travelling is related to the idea of being simultaneously on the move and at the destination. It is a part of the very being of transparency. To be here and somewhere else at the same time. Transparency opens up both sides. Together steel and glass create versatile possibilities to realise the illusion.

The lightness that can be achieved by steel and glass is one of the challenges of this millennium. Buckminster Fuller always emphasized the weight of a building. According to the new ecological way of thinking, this also involves the energy consumed during the construction process and the energy consumed to maintain the building.

In traffic architecture the challenges of the 3rd millennium could be lightness, dynamics, transparency, light, speed, precision and credibility.

When travelling, however, the most important thing is to be able to find your next destination as easily as possible. That is why signs are part of traffic architecture. The more the space and environment itself guide you the better. Signs should enhance the visual quality of the space. Waiting is part of the journey. The comfort of waiting has not been paid enough attention. The principles of nest building are important when waiting is concerned.

In the actual transport vehicle, both basics are emphasized. The vehicle shall reflect dynamics and provide comfort at the same time. This means that the desing of transport vehicles faces ever growing challenges.

欧洲所有的大型火车站中，分布着很多的站台。

既然我们已经进入了新的千年，赫尔辛基火车站的站台庭院将最终就像它被期待的那样得到保护。这些交通建筑设施工程致力于使旅行变得更加舒适，特别是在我们国家北部地区有时非常令人难受的气候里。它们也成为在交通建筑设施的发展过程中，在透明性以及使用钢和玻璃的重要性方面提供的一些个案研究。

工程：

1 Seinä joki 火车站　1993年

2 Kaisaniemi Metro 火车站　1995 年

3 Pori 火车站　1998 年

4 Vuosaari Metro 火车站　1998 年

5 Korso 火车站　2000 年

6 Helsinki 火车站站台雨棚　2001 年

The vigorous development of railways towards the end of the 19th century was reflected in a boom in the construction of railway stations. Platforms were covered at all the main railway stations in Europe.

And now that we have entered the new millennium, the station yard at Helsinki Railway Station will finally get the shelters that have been waited for.

These projects of traffic architecture are efforts to make travelling more comfortable in the sometimes quite harsh climate of our northern country.

They are also case studies in the nature of transparency and the significance of steel and glass in the development of traffic architecture.

Projects:

1 Seinäjoki Railway Station 1993

2 Kaisaniemi Metro Station 1995

3 Pori Railway Station 1998

4 Vuosaari Metro Station 1998

5 Korso Railway Station 2000

6 Helsinki Railway Station Platform Roofing 2001

1

2

3

4

5

6

Curriculum vitae

Esa Piironen
Professor, Helsinki

履历

埃萨 · 皮罗宁，芬兰建筑师协会会员，美国建筑师学会荣誉会员，
教授，赫尔辛基

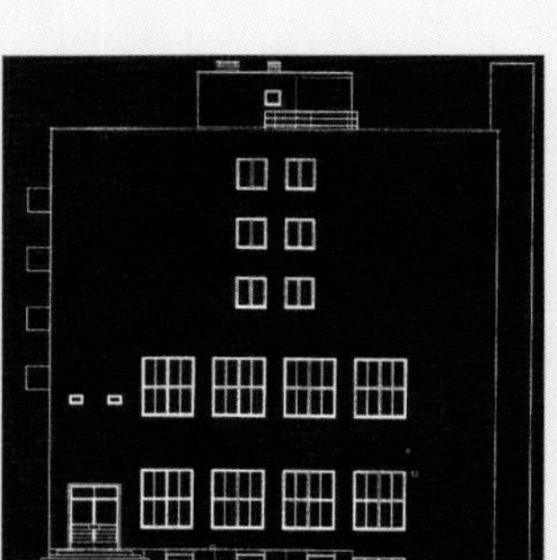

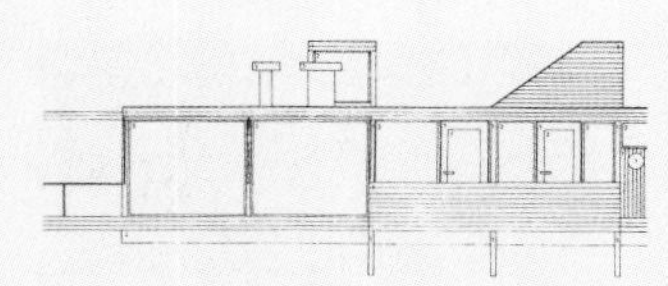

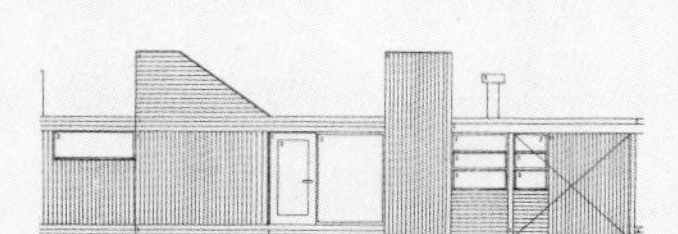

EDUCATION

Licentiate in Technology, Helsinki University of Technology, Otaniemi	1978
Master of Architecture, N.C.S.U., USA	1972
Architect, Helsinki University of Technology, Otaniemi	1970

TEACHING EXPERIENCE

Architecture 1, senior assistant, Helsinki University of Technology	1972-81
(professor Martti Jaatinen): practical works programming and lectures once a year on architecture, presentation techniques and environmental psychology Landscape design, assistant, Helsinki University of Technology	1970-71
Environment design, teacher, Institute of Industrial Arts, Helsinki	1970-71
Lectures on various subjects related to urban design, architecture, graphic design, exhibition design: North Carolina State University, School of Design, Raleigh (prof. Peter Batchelor)	1972
INSKO, Yyteri, Urban design and environmental psychology	1976
Helsinki University of Technology, Department of Architecture, Otaniemi (professor Martti Jaatinen) once a year	1981-91
University of Washington, Seattle (professor Phillip L. Jacobson)	1982
Kouvola School of Industrial Arts	1992
Tampere University of Technology, Department of Architecture (steel construction course)	1995
Oulu University, Department of Architecture (advanced steel construction course)	1997
Tampere University of Technology, Department of Architecture (professor Reijo Jallinoja)	2001

EMPLOYMENT

Architectural Office Esa Piironen Oy, Helsinki	1990-
Architectural Office Sakari Aartelo and Esa Piironen, Helsinki	1978-90
Architectural Office G4, Helsinki, partner	1970-85
Espoo City, Town Planning Office	1970-72
Architectural Office Esa Piironen and Mikko Pulkkinen, Turku	1966-71
Architectural Office Reino Aarnio, New York	1964
Architectural Office Pekka Pitkänen, Turku	1963-69
Architectural Office Pekka Pitkänen and Olli Vahtera, Turku	1959-62

MEMBERSHIPS AND POSITIONS OF TRUST

OISTAT Finland, Board of Directors	2001-
Member of the Register of Architects in Finland, ARK 257	2000
Association of Architectural Offices, Helsinki	1991-
Chairman of the Competition Committee, Finnish Association of Architects	1990-91
Graphic Designers, Helsinki	1978-
Grafia, Helsinki	1976
Helsinki University of Technology, Department of Architecture, College of Staff	1973,1975-76
EDRA, Environmental Design Research Association	1972
NCSU, Alumni	1972
ASLA, Alumni	1972
Association of Planners and Urban Designers, Helsinki	1970
SAFA, Finnish Association of Architects	1970-
International Center for Typographic Art, New York	1968
Jyväskylä Cultural Days, Member of the visual committee	1967
Chairman of the Architectural Student Club, Otaniemi	1966-67

SCHOLARSHIPS AND GRANTS

State artist grant	1999-2003
Residence scholarship, Cité Internationale des Arts, Paris	1999
Residence scholarship, Center for Visual Arts, New York	1994
Residence scholarship, Cité Internationale des Arts, Paris	1989
Greta and William Lehtinen Foundation, grant	1978
Finnish Academy, travel grant, EDRA, USA	1977
Väinö Vähäkallio grant	1975
Lasse and Kate Björk Foundation grant	1973
ASLA-Fulbright, USA	1971-72
Finnish Cultural Foundation, Varsinais-Suomi provincial grant	1968
Oy Ulkomainos Ab, scholarship	1967

JURY MEMBERSHIPS

Viikki Triangle cemetery sculpture	1999
Aurinkolahti Secondary School, Helsinki	1999
Vuosaari subway station, art competition	1996
Finnish Association of Architects, competition for the centenary emblem	1991
Gateway to Hämeenlinna, invited competition	1990
Tampere Hall, art competition	1989
Center for Allergic Rehabilitation, Jalasjärvi, invited competition	1989
Finnish Science Center, Vantaa	1985
Kiuruvesi Town Hall	1980
Museum of Finnish Architecture, exhibition design competition	1967

AWARDS

Pormestari pedestrian bridge, Pori, 1st prize	2000
Embassy of Finland. Canberra, honorary mention	1997
Pori railway-station, pedestrian tunnel and platform shelter, 1st prize	1996
Helsinki railway station, platform roofing, 1st prize	1995
A type-kiosk for Helsinki, 3rd prize	1994
Luuniemi Housing Area, Iisalmi, 1st prize	1993
Västra Eriksberg Housing Area, Göteborg, honorary mention with others	1992
Tikkurila Orthodox Church, 3rd prize	1991
Street Furniture, Espoo, 2nd prize	1991
Lutakko Housing Area, Jyväskylä, 3rd prize, assistant	1990
Viikki Triangle cemetery, invited competition, 1st prize, A+P	1987
Lamminpää cemetery and chapel, invited competition, honorary mention., A+P	1986
Laajasalo Church, invited competition, 1st prize, A+P	1984
The Official Residence of the President of Finland, 2nd prize, A+P	1983
Tampere Concert and Congress Hall, 1st prize, A+P	1983
Järvenpää Administrative and Cultural Center, purchase, A+P	1981
Rauma Town Hall, purchase, A+P	1981
Pori Music and Congress Hall, 3rd prize, A+P	1981
Iisalmi Library and Cultural Center, 3rd prize, A+P	1980
Rautavaara Church, invited competition, 1st prize. A+P	1979
Kouvola Cultural Center, 2nd prize, A+P	1979
Multia Main Village, 3rd prize, A+P	1979
Kumpula University Area. Helsinki, 1st prize (shared), A+P	1978
Tampere City Library, purchase, A+P	1978
Pyhäjoki Church, 1st prize	1975
Tampere Labor Theater, invited competition, 1st prize assistant	1974
Puolivälinkangas Church, Oulu, 2nd prize	1970
Lounaisrannikko Terraced Houses, Espoo, 3rd prize with	

others 1968
Art Festival, Taidetapahtuma 1, Turku, poster competition, 1st prize 1966
Otaniemi Finland, poster competition, 1st prize 1966

HONOUR AWARDS

Brunel Award 2001, honorary mention 2001
Steel Construction Award, A+P 1986
Poster among the best yearly posters in Finland 1964

SELECTED PROJECTS

Mäntsälä railway station 2003-
Haarajoki railway station, Järvenpää 2003-
Korsokeskus pedestrian bridge, Vantaa 2003
Mikael Agricola Church, renovation, Helsinki 2002-
Urpia railway station, Vantaa 2002-
Mikkola School, Vantaa 2002-
Highway 4, Lahti-Heinola, design 2002
Rautaruukki Polska, Technology Center, Zyrardow 2002
Leppävaara Exchange Terminal, Espoo 2002
Kehä 1 traffic noise barriers, Kannelmäki-Lassila, Helsinki 2002
Nöykkiönlaakso school annex and reparation, Espoo 2002
Latokaski-Nöykkiö health center, Espoo 2002
Nöykkiö library and youth club annex and reparation, Espoo 2002
Hiekkaharju railway station, Vantaa 2001-
Koivukylä railway station, Vantaa 2001-
Roofing Helsinki Pedestrian Streets, plan 2001
Helsinki railway station, platform roofing 2001
Pormestari pedestrian bridge, Pori 2001
Pikku-Huopalahti elementary school annex, Helsinki, project 2001
Two-family house, Espoo 2000
Korso railway station, Vantaa 2000
Rekola railway station reparation, Vantaa 2000
Nöykkiö day care center reparation, Espoo 2000
Kallio Municipal Offices reparation, Helsinki 1999-
Helsinki 2nd Training School reparation and annex 1999
Vuosaari subway station, Helsinki 1998
Pori railway station, pedestrian tunnel and platform shelters 1998
Pikku-Huopalahti Multipurpose House, Helsinki 1997
VR, Reparation program for the local stations, Helsinki 1996
VR, Koivuhovi railway station 1996
Kaisaniemi subway station, platform hall, Helsinki 1995
Tapanila Church reparation, Helsinki 1995
HKL, Rautatientori Ticket Office, Helsinki 1994
Seinäjoki railway station, pedestrian tunnel and platform shelters 1993
Hietaniemi cemetery, Helsinki, lighting fixtures 1993
Bus shelter, Public Design '92, Espoo 1992
Kalevantie 16, Turku, Office building, project 1992
Kauhajoki School of Domestic Economics, reparation and annex 1992
Piispanhelmi, Espoo, Office building, project 1991
Tiistinportti, Espoo, Office building, project 1991
Viikki Triangle cemetery, Helsinki, together with Leena Iisakkila, A+P 1990
Office building Rautio, Espoo, A+P 1990
Tampere Hall, A+P 1990
Piispankylä 22286, Espoo, Area plan 1990
Atrium, Espoo, Office building, project 1990
Kurkimäki Multipurpose House, Helsinki, A+P 1989
Äänekoski cemetery, together with Leena Iisakkila, A+P 1989
Helsinki University, Kumpula Chemistry building, invited competition, A+P 1988
Private House Koivikko, Rohkatie 20, Helsinki 1985
Forest Research Center, Kannus, A+P 1985
Botanical Garden and Glass House, Joensuu University, A+P 1985
Hansasilta Shopping Mall and Pedestrian Bridge, Itäkeskus, Helsinki, A+P 1984
Rautavaara Church, A+P 1982
Kallio School restoration, Helsinki, A+P 1982
Veräjämäki Parish Center, Helsinki, project, A+P 1981
Private House Agge, Parainen 1980
Private House Kokkala, Helsinki 1980
Row Houses, Kielokallio, Espoo, rivitalo 1978
Sauna Färm, Pakinainen, project 1975
Nuottaniemi 1, Espoo, Town Plan, G4 1975
Villa Björköholmen, Parainen, annex 1974
Tiistilä, Espoo, Town Plan, G4 1973
Weekend Cottage Pöllö, Velkua 1971
Villa Vainiola, Aura 1970
Villa Suojararanta, Merimasku 1969
Grillkiosk 18, Turku, demolished 1969
Private House Sunila, annex, Piikkiö, P+P 1968
Lomatalo, House system, project, P+P 1968
Private House Viljanen, Turku, project, P+P 1968
Private Swimming Hall, Turku, P+P, demolished 1967
Villa Troberg, Förby, project, P+P 1967
Chapel of Holy Cross, Turku, assistant by architect Pekka Pitkänen 1967
General Plan, Hirvensalo, Turku, project, P+S 1966
Private House, Söörmarkku, P+P 1966
Villa and sauna Laakso, Kangasala, P+P 1966
Exhibition pavillion, Turku Fair, P+P 1966
Sauna Suojaranta, Merimasku 1964

EXHIBITION DESIGN

Steel Construction Awards 1980-2000, TRY, Finlandia Hall, Helsinki 2001
Marc Chagall et.al., Tampere Hall 1992
Lars Sonck 1870-1956, A Masieri Fondation, Venice 1990
Lars Sonck 1870-1956, Konstakademien, Stockholm. Konsthistoriska museet, Göteborg 1989
Ars Sacra Fennica, Kemi Art Museum; Joensuu Art Museum 1989
Tampere Hall 1990, Sara Hildén Art Museum. Tampere, A+P 1988
Ars Sacra Fennica, Galerie für Christliche Kunst, München; Rovaniemi Art Museum 1988
Ars Sacra Fennica, Helsinki 1987
Lars Sonck 1870-1956, Imatra 1987
Lars Sonck 1870-1956, Tampere Art Museum 1986
Lars Sonck 1870-1956, Tornionlaakso; Uusikaupunki; Hanko; Rauma, Varkaus 1985
Lars Sonck 1870-1956, AIA, Washington DC; University Gallery, Minneapolis; National Academy of Design, New York; Art Institute of Chicago 1983
Lars Sonck 1870-1956, Heinz Gallery, London; Oulu University; Gould Hall, University of Washington, Seattle 1982
Lars Sonck 1870-1956, Museum of Finnish Architecture, Helsinki; Aalto Museum, Jyväskylä; Järvenpää 1981
G4, imago, Lahti Poster Museum 1978
G4, imago, Turku Art Museum 1978
Street Furniture. Rauma Museum, G4 1977
G4, imago, Kluuvi Art Gallery, Helsinki 1977
Street Furniture, Vantaa, G4 1976
Street Furniture, Oulu University, G4 1976
Street Furniture, Johannes Haapasalo Museum, Mikkeli, G4 1975
Old Porvoo, Porvoo 1971
Street Gallery and Furniture, Turku Art Museum,G4 1971
Opi metro, A sign system for Helsinki subway, G4 1971

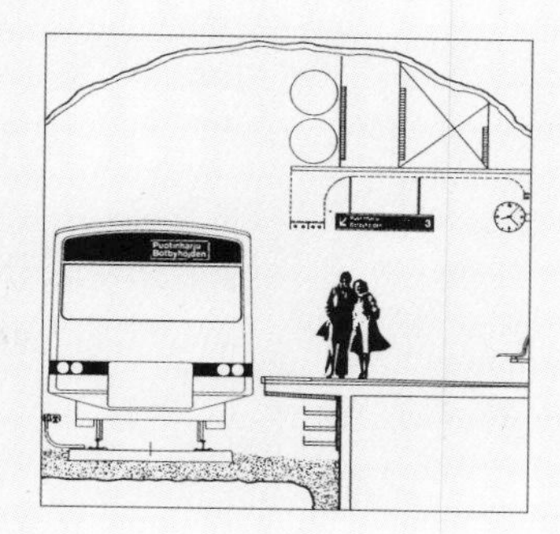

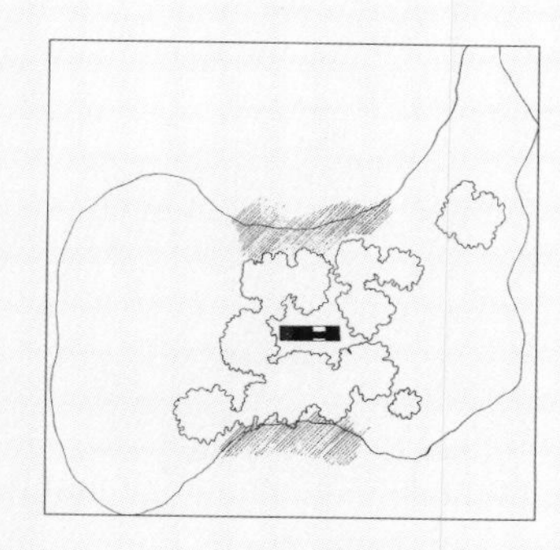

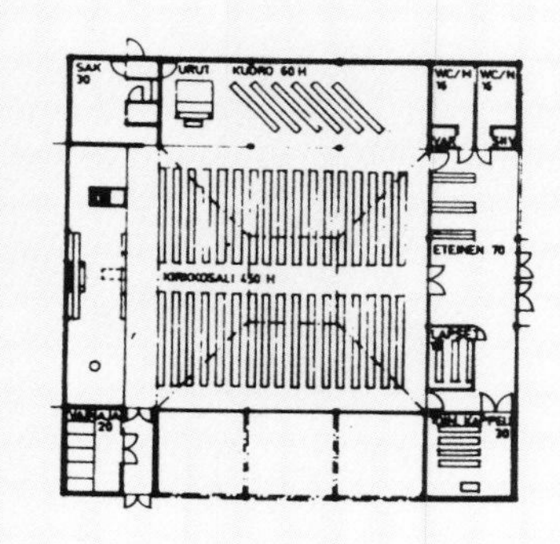

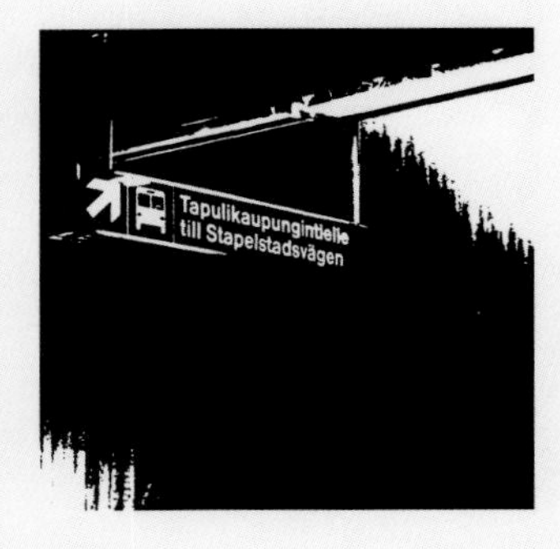

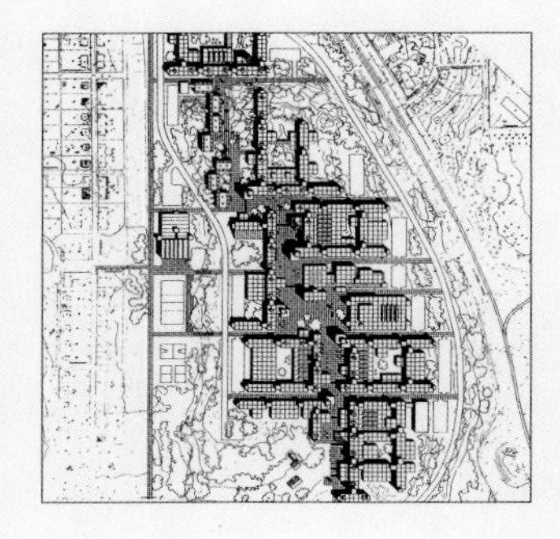

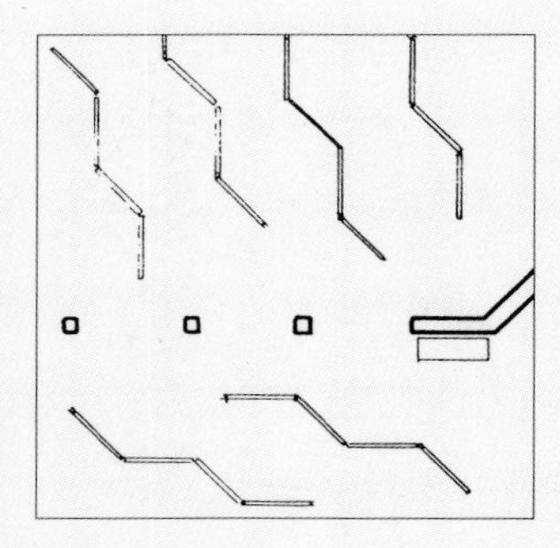

- Street Furniture, Museum of Finnish Architecture, G4 — 1970
- Old and New Espoo, Dipoli, Otaniemi — 1970
- Street Gallery, International Poster Exhibition, Taidehalli, Helsinki, G4 — 1969
- Erik Bryggman 1891-1955, Museum of Finnish Architecture, Helsinki — 1968
- Graphics, Teekkarigrafiikka, THS Studentkår, Stockholm, G4 — 1967
- Graphics, Teekkarigrafiikka, Dipoli, Otaniemi, G4 — 1967
- FEPE, Poster Exhibition, Dipoli, Otaniemi — 1967
- Erik Bryggman 1891-1955, Turku Art Museum — 1967
- Art Festival, Taidetapahtuma 1, Turku Art Museum — 1966

SIGN SYSTEM DESIGN

- Otaniemi sign system, Espoo — 2002
- Helsinki Trade Fair Center — 1999
- Helsinki subway, sign system design instructions — 1997
- Helsinki center terminal sign system — 1997
- Helsinki-Vantaa Airport sign system, basic concept — 1994
- Helsinki railway station sign system — 1994-95
- Finnish State Railways, local stations (Pasila-Hiekkaharju), sign system — 1992
- Kaisaniemi subway, station, Helsinki — 1991-95
- Kauhajoki School of Domestic Economics — 1990
- Tampere Hall, A+P — 1990
- Aurora Hospital, Helsinki, G4 — 1984
- University of Lappland, Rovaniemi, G4 — 1984
- Laakso Hospital, Helsinki,:G4 — 1982
- Orion Medical Center, Espoo, G4 — 1981
- Riihimäki Swimming Hall, G4 — 1979
- Varissuo Housing Area, Turku, G4 — 1978
- Turku University, G4 — 1978
- Helsinki subway, sign system, realization plan, G4 — 1977-82
- Finnish State Railways, passenger sign system, G4 — 1976-82
- Joensuu University, G4 — 1976
- Martinlaakso Line Stations, G4 — 1975
- Kuopio University, G4 — 1975
- Helsinki subway, sign system, general plan, G4 — 1975
- Merihaka Housing Area, Helsinki, G4 — 1972
- Valio Dairy Plants, G4 — 1972
- Helsinki subway, sign system, preliminary plan, G4 — 1972
- A sign system, NCSU — 1971
- Turku Student Village — 1971
- Helsinki subway, sign system, report, G4 — 1970

GRAPHIC DESIGN

- Teräs julkisessa rakentamisessa, graphic design — 1998
- YTV, public transport maps, graphic design — 1995
- Finnish State Railways, route map and time tables, graphic design — 1994
- Marc Chagall et.al., Tampere Hall, poster and graphic design — 1992
- Takstooli-magazine, graphic design — 1991
- Novaco Ltd, trade mark, together with Jyrki Nieminen — 1990
- Orion Ltd, Espoo, graphic design — 1989
- Ars Sacra Fennica, poster and graphic design, together with Jyrki Nieminen — 1987
- Abacus 3, Museum of Finnish Architecture, graphic design — 1983
- Finnish State Railways, sign system, graphic design — 1982
- Rakennushallitus, graphic design — 1981
- Lars Sonck, 1870-1956, posters and graphic design — 1981-83
- Helsinki University of Technology, graphic design — 1980
- Finnish Cinema Archives, graphic design — 1980
- Valmet Ltd, corporate design, project — 1979
- Marli Ltd, Turku, corporate design — 1978
- F.O.Bergman Ltd, Turku, corporate design — 1978
- Finnish Academy, graphic design — 1978
- G4, imago, graphic design — 1977
- Sports and Leisure, graphic design — 1977
- Helsinki subway corporate design 2880 SAT 2, graphic design — 1976
- Helsinki subway sign system, graphic design — 1975
- Helsinki subway design hand book, graphic design — 1974
- Department of Architecture, Helsinki University of Technology, graphic design — 1973
- Building Information Foundation, graphic design — 1972
- Ikituuri Hotel, Turku, corporate design — 1971
- Opi metro-exhibition, G4, poster and graphic design — 1971
- Espoo City Town Planning Office, graphic design — 1970
- Bureau of Standardization, SAFA, graphic design — 1970
- Street Furniture, G4, poster and graphic design — 1970
- Espoo Local Festival, posters and graphic design — 1970
- City Research 70, graphic design — 1970
- Espoo City Plan 1970-80, graphic design — 1969
- Turku University Student Union, graphic design — 1969
- Turku General Plan Report, graphic design — 1969
- Street Gallery, poster exhibition, G4, graphic design — 1969
- a-magazine, Arkkitehtikilta, graphic design — 1968
- Tavastin Kilta Restaurant, Naantali, corporate design — 1968
- Länsi-Meri Restaurant, Rauma, corporate design — 1968
- Summerhotel Studenthouse, Turku, corporate design — 1968-70
- Commentationes Physico-Mathematicae, Helsinki University, graphic design — 1968
- Vaasa Summer University, graphic design — 1968
- Graphics, Teekkarigrafiikka, G4, graphic design — 1967
- Erik Bryggman 1891-1955, poster and graphic design — 1967
- Finland in Focus, Contactor, graphic design — 1967
- Helsinki University Student Union, poster — 1967
- Projektio-magazine, graphic design — 1967
- FEPE, General Assembly, Helsinki, poster and graphic design — 1967
- Turun Ylioppilaslehti-magazine, graphic design — 1967-68
- Jyväskylä Cultural Days, graphic design together with Ola Laiho — 1964
- Helsinki University of Technology Student Union, Espoo, graphic design — 1964
- Restaurant Dipoli, Espoo, graphic design — 1964
- Art Festival Taidetapahtuma 1, Turku, poster and graphic design — 1964
- Tekniikan Suomi maailmankartalle, TKY, poster together with Esko Miettinen — 1964
- Teekkari-magazine, graphic design — 1964-67

PUBLICATIONS

- Small Houses in Finland, Rakennustieto Oy, Helsinki (in preparation) — 2003
- Monograph: Esa Piironen, Architect, China Architecture and Building Press (in preparation) — 2003
- Roofing Helsinki Pedestrian Streets, HKR — 2001
- Steel Images, TRY, Rakennustieto Oy, Helsinki, editor — 2001
- Teräs julkisessa rakentamisessa, TRY, Rakennustieto Oy, Helsinki together with Risto Saarni — 1998
- Helsingin metron opastusjärjestelmän suunnitteluohje, HKL — 1997
- Kaupunkirata Helsinki-Huopalahti-Leppävaara, asemasuunnittelun ohjeet, Ratahallintokeskus, Helsinki — 1997
- Tikkurilan joukkoliikennepysäkki, Vantaa, C5:1996 — 1996
- Helsingin keskustan terminaalien matkustajaninformaation kehittäminen, HKL et.al., Helsinki — 1995
- Helsinki Vantaan lentoaseman opastesuunnittelun perusteet, Helsinki, together with Viisikko Oy — 1994
- Ulkomainonta kaupunkikuvassa, FEPE Finland, Helsinki, editor — 1983

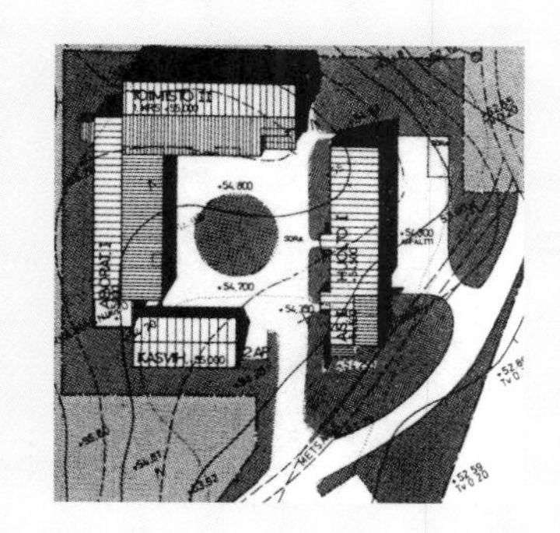

VR, Opastusjärjestelmä, Helsinki 1982
Lars Sonck 1870-1956, Museum of Finnish Architecture, Helsinki, editor 1981
Kurkimäen korttelitalo, Esisuunnitelma. HKR, Helsinki, A+P 1980
Suomalaisia kytkettyjä pientaloja, TKK, Espoo, editor 1980
Le Corbusier: Uutta arkkitehtuuria kohti, translation, TKK, Espoo, editor 1979
Matalaenergiataloja, TKK, Espoo, editor 1978
Ympäristöpsykologia 5, TKK, Espoo, editor 1978
Ympäristön havaitsemisesta ja sen mittaamisesta, TKK, Espoo 1978
Suomalaisia yhdenperheentaloja, TKK, Espoo, editor 1977
G4, imago, Helsinki 1977
Sports and Leisure, Helsinki, editor 1977
Ympäristöpsykologia 4, TKK, Espoo, editor 1977
Ympäristöpsykologia 3, TKK, Espoo, editor 1976
Suomalaisia loma-asuntoja ja saunoja, TKK, Espoo, editor 1976
Helsingin metron opastus- ja informaatiojärjestelmä, yleissuunnitelma, G4 1975
Ympäristöpsykologia 2, TKK, Espoo, editor 1975
Metroasemien sisustuskomponentit, Helsinki, editor with others 1974
Lastentalon suunnittelun perusteet, TKK, Espoo, editor 1974
Loma-asunnon suunnittelun perusteet, TKK, Espoo, editor 1974
Ympäristöpsykologia 1, TKK, Espoo, editor 1974
Metro/USA, Helsingin kaupungin metrotoimisto 1974
TKK/A/100, Arkkitehtiosaston juhlajulkaisu, editor 1974
Arkkitehtuuriopas Turku, Arkkitehti-magazine 1972
Helsingin metron opastus- ja informaatiojärjestelmä, G4 1972
Kadun kalusteet, Museum of Finnish Architecture, Helsinki, G4 1970
Helsingin metron opastus- ja informaatiojärjestelmätutkimus, G4 1970
Tutkimus seutu- ja yleiskaavamerkinnöistä, SAFA Asemakaava- ja standardisoimislaitos, Helsinki 1970
Erik Bryggman 1891-1955, Turku, editor 1967
Teekkari-magazine, editor, TKY 1964-66

ARTICLES

Kerava City Rail Line Tikkurila-Kerava, Teräsrakenne 1/2003 2003
Rautaruukki Polska, technology centre, Teräsrakenne 4/2002 2002
Leppävaara interchange terminal, Teräsrakenne 3/2002, Helsinki 2002
Katettu katu, Teräsrakenne 2/2002, Helsinki 2002
Korso railway station, Teräsrakenne 2/2002, Helsinki 2002
Roofing of Helsinki Railway Station Yard, area 2, Teräsrakenne 4/2001 2001
Steel and glass texture, Arkkitehti 5/2001, Helsinki 2001
Pormestarinsilta Overpass, Pori, Teräsrakenne 3/2001, Helsinki 2001
Eliel Saarinen's Roofing Plans for Helsinki Railway Station Yard, Teräsrakenne 2/2001, Helsinki 2001
Glass and Steel in Traffic Architecture, Glass Processing Days, Tampere 2001
Kaksoislasijulkisivuarkkitehtuurista, Teräsrakenne 1/2001, Helsinki 2001
Millenniumarkkitehtuuria, Teräsrakenne 4/2000, Helsinki 2000
Jazzarkkitehtuuria, Teräsrakenne 2/2000, Helsinki 2000
Roofing of Helsinki Railway Station Yard, Teräsrakenne 3/99, Helsinki 1999
Portal Tunnel in Pori, Teräsrakenne 4/98, Helsinki 1998
Teräs julkisessa rakentamisessa, HKR-Rakennuttaja, Helsinki 1998
Bilbaon ruusu, Teräsrakenne 1/98, Helsinki 1998
Helsinki Railway Station Platform Roofing, Teräsrakenne 1/98, Helsinki 1998
Pikku-Huopalahti Block Building, Teräsrakenne 4/97, Helsinki 1997
Vuosaari Subway Station, Teräsrakenne 2/97, Helsinki 1997
Pikku-Huopalahti Block Building, Teräsrakenne 2/97, Helsinki 1997
Seinäjoki Station Underpass and Platform Canopies, Arkkitehti 4/96, Helsinki 1996
Liikenteen teräsarkkitehtuurista, Teräsrakenne 1/96 1996
Helsingin rautateiaseman kattamiskilpailu, Teräsrakentaminen/ TTKK, Tampere 1995
Kaisaniemi Metro Station platform area, Teräsrakenne 3/95, Helsinki 1995
Seinäjoki Station Pasenger Tunnel and Platform Shelters, Teräsrakenne 1/94 1994
Talo Koivikko, Arkkitehti 7-8/93, Helsinki 1993
Terästä auringossa, Teräsrakenne 2/92 1992
Talo kuin kylä, Kurkimäen korttelitalo, Puu 2/91, Helsinki 1991
Kauhajoen koti-ja laitostalousoppilaitos, Takstooli 3/91, Helsinki 1991
Down under-terästä Sydneyssä, Teräsrakenne 1/91 1991
Small is beautiful, Erik Bryggman's villa architecture, Museum of Finnish Architecture 1991
Stålbyggandets tradition fortsätter i Paris, Bygg & Fastighets Utveckling 3/90 1990
Kurkimäen korttelitalo, Tiili 3/90, Helsinki 1990
Pariisin teräsrakentamisen perinne jatkuu, Teräsrakenne 1/90 1990
Rautatiematkustajien opastusjärjestelmä, Rautatieiiikenne 1/80 1980
Designing a sign system, Arkkitehti 7/78 1978
Measuring human responses to the environment, HTKK 1978
Opastusjärjestelmän suunnittelu, G4, imago 1977
Suomalaisia urheilurakennuksia, Arkkitehti 4/77 1977
Ympäristön havaitsemisen mittaamisesta, Ympäristöpsykolgia 4, HTKK 1977
Kadun kalusteet, Arkkitehti 8/70 1970
Ulkomaat GB, Arkkitehti 7/69 1969
Ulkomaat 4, Arkkitehti 4/69 1969
Ulkomaat 2, Arkkitehti 2/69 1969
Ulkomaat 1, Arkkitehti 1/69 1969
Monimutkainen ja ristiriitainen arkkitehtuuri, Arkkitehti 5/68 1968
Käyttögrafiikkaa G4, Arkkitehti 1-2/67 1967
Erik Bryggman 1891-1955, Suomen Turku 1967
Hommage á Pablo Picasso, Turun Ylioppilaslehti 1967
Hirvensalon yleiskaava, Taidetapahtuma 1, Turku, P+S 1966
G4, Teekkarigrafiikkaa, Taidetapahtuma 1, Turku 1966
A 2000, Johdatus tulevaisuuden asumiseen, Taidetapahtuma 1, Turku 1966
Mielentiloja vapaudesta, amerikkalaista grafiikkaa Turussa, Aamulehti 1965
Amerikkalaisesta arkkitehtikoulutuksesta, a-lehti 1965
Amerikkalaisista arkkitehtitoimistoista, a-lehti 1965
Pelouze, amerikkalaisista arkkitehtitoimistoista, Teekkari 3/64 1964

WORKS IN EXHIBITIONS

Contemporary Urban Architecture in Helsinki, Helsinki 2001
8th Brunel Awards International Railway Design Competition, Paris 2001
Steel Construction Awards 1980-2000, TRY, Finlandia Hall, Helsinki 2001
Suomi Viherrakentaa 2001, Helsinki 2001
Kaupunkisuunnittelu 2000, Helsinki 2000
Rautaiset rakenteet, Museum of Finnish Architecture, Helsinki 1998
Tampere-exhibition, Chemnitz, Linz, Essen, Lodz 1996
Rakennettu puusta, Museum of Finnish Architecture,

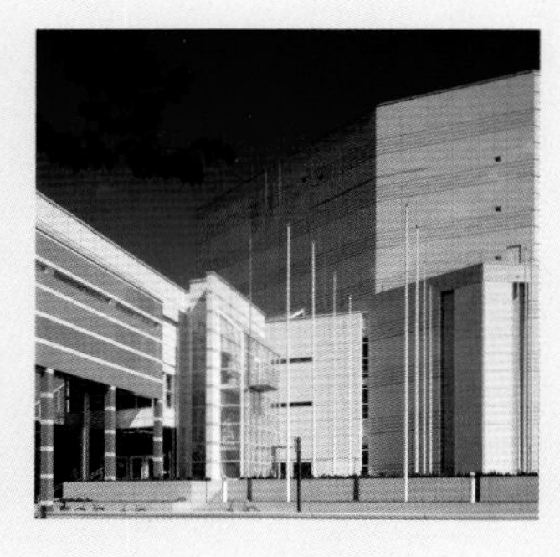

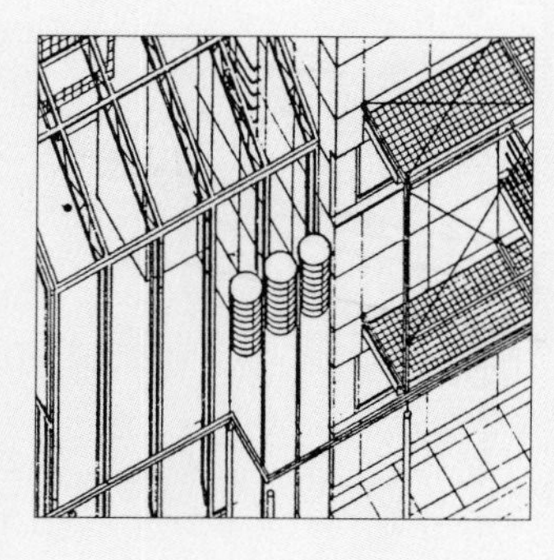

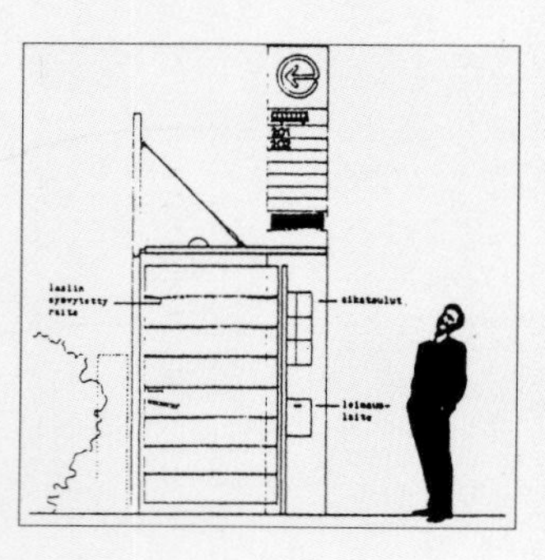

Helsinki 1996
Tampere-exhibition, Mini Europa, Brussels 1995
Tampere-näyttely, Finnish Institute, Paris 1995
Suomalaista arkkitehtuuria, Technical University, Warsaw 1994
Suomi muotoilee 9, Helsinki 1993
100+, Suomalaisia liikennemerkkejä, Helsinki 1992
Suomi rakentaa 8, Helsinki 1992
Prague Quadriennale 1991
Tampere-talo 1990, Sara Hilden Art Museum, Tampere 1988
Ars Sacra Fennica, Helsinki, Munich 1987
Suomalainen pientalo, Helsinki 1986
Rakennushallitus 175-vuotta, Helsinki 1986
Suomi muotoilee 4, Helsinki 1983
Suomalaista käyttögrafiikkaa, Musashino Art Institute, Japan 1983
Vuoden huiput, Helsinki 1981
Suomi muotoilee 3, Helsinki 1979
MG-25-vuotta, Finlandia Hall, Helsinki 1978
G4, imago, Helsinki 1977
Suomi rakentaa 4, Helsinki 1970
Venice Biennale 1969
Teekkarigrafiikkaa G4, Dipoli, Espoo 1967
Taidetapahtuma 1, Turku 1966
Vuoden parhaat julisteet, Helsinki 1964

WORKS IN PUBLICATIONS

Olli Helen: Sadan vuoden urakka, Karisto, Hämeenlinna 2003
Teräsrakenne 1/2003, Helsinki 2003
ATL, Arkkitehtitoimistot 2003-2005, Vammala 2003
Teräsrakenne 4/2002, Helsinki 2002
100 of the World's Best Houses, Images Publishing, Mulgrave, Australia 2002
Teräsrakenne 3/2002, Helsinki 2002
Finnish New Architecture, ed. Fang Hai, China 2002
Talotekniikka 2/2002, Helsinki 2002
Rakennustiedon historiikki, Rakennustieto Oy, Helsinki 2002
Teräsrakenne 2/2002, Helsinki 2002
Timo Koho: Heurekasta Kiasmaan, Rakennustieto Oy, Helsinki 2002
L'ARCA 165, Milan 2001
Matti Rinne: Aseman kello löi kolme kertaa, Otava, Keuruu 2001
Details in Architecture 3, Images Publishing, Mulgrave, Australia 2001
Teräsrakenne 4/2001, Helsinki 2001
Arkkitehti 5/2001, Helsinki 2001
8th Brunel Awards International Railway Design Competition, Paris 2001
Water Spaces of the World, Images Publishing Group, Melbourne 2001
Projektiuutiset 4/2001, Espoo 2001
Little Big Houses, Building Information Ltd, Helsinki 2001
Teräsrakenne 3/2001, Helsinki 2001
Contemporary Urban Architecture in Helsinki, Rakennustieto Oy, Helsinki 2001
Architecture Today, Nro 120, July, 2001, London 2001
Glass Processing Days, Tampere 2001
Steel Images, TRY, Rakennustieto Oy, Helsinki 2001
Finnish Landscape Architecture 2001, Viherympäristöliitto, Helsinki 2001
Päivi Lampinen: Miten ja miksi koulurakennus muuttuu?, Opetusvirasto, Helsinki 2000
ATL, Arkkitehtitoimistot 2000-2002, Vammala 2000
Details in Architecture 2, Images Publishing, Mulgrave, Australia 2000
Hej! 6/2000, IKEA, Vösendorf, Austria 2000
Forum för ekonomi och teknik 11/2000, Helsinki 2000
Teräsrakenne 3/2000, Helsinki 2000
Lasirakentaja 2/2000, Tampere 2000
The World of Contemporary Architecture, Könemann mbH, Cologne 2000
Arkkitehtuurin sanakirja, WSOY, Helsinki 2000
Insight Guide Finland, Apa Publications, Singapore 2000
Leena Iisakkila: Maisema-arkkitehti ajan virrassa, Forssa 2000
Lasirakentaja 1/2000, Tampere 2000
Innovations in Steel, Transparent Architecture, IISI, Rotterdam 1999
Jorma Mukala: Tampereen arkkitehtuuria 1999
Teräs pientalorakentamisessa, TRY, Rakennustieto Oy, Helsinki 1999
Finnish Architecture 1994-1999, Photographed by Jussi Tiainen, Helsinki 1999
Teräsrakenne 3/99, Helsinki 1999
International Architecture Yearbook No. 5, Mulgrave, Australia 1999
Teräsrakenne 4/98, Helsinki 1998
Projektiuutiset 4/98, Helsinki 1998
Teräsrakenne 3/98, Helsinki 1998
Teräs julkisessa rakentamisessa, TRY, Rakennustieto Oy, Helsinki 1998
Teräsrakenne 2/98, Helsinki 1998
Rakennuslehti Nro27/17.9.1998, Helsinki 1998
Sähköala 5/98, Espoo 1998
Tubular Structures 68, Railway Stations and Depots, John Pascoe, toim., Corby 1998
Rautaiset rakenteet, Museum of Finnish Architecture, Helsinki 1998
Teräsrakenne 1/98, Helsinki 1998
Arkkitehtuurikilpailuja 1/98, Helsnki 1998
ATL, Arkkitehtitoimistot 1997-98, Vammala 1997
Kaipia-Putkonen: Suomen arkkitehtuuriopas, Otava, Keuruu 1997
Helsinki/City in the Forest, Ichigaya Publishing Co., Tokyo 1997
Teräsrakenne 4/97, Helsinki 1997
Projektiuutiset 4/97, Helsinki 1997
Rakennustekniikka 2/97, Helsinki 1997
The Fourth Wave of Rock Construction, Porvoo 1997
Teräsrakenne 2/97, Helsinki 1997
Meidän Talo 6/97, Helsinki 1997
Pikku Huopalahden korttelitalo: tuoteosakauppa, paikallarakennettavat teräsrankaiset ulkoseinärakenteet: Ari Roininen, thesis, TTKK, Tampere 1997
Le Carré Bleu 3-4/96, Paris 1996
Teräsrakenne 4/96, Helsinki 1996
aw 168, architektur+wettbewerbe, Stuttgart 1996
Arkkitehti 4/96, Helsinki 1996
Rakennettu puusta, Suomen rakennustaiteen museo, Helsinki 1996
Competition & Contest 9601, Tokyo 1996
Helsingin rautatieasema, Helsinki 1996
Rakennustaito 5/95, Helsinki 1995
Lasirakentaja 2/95, Tampere 1995
Teräsrakentaminen, TTKK, Tampere 1995
Valo 1/95, Helsinki 1995
Werk, B+W 10/95, Munich 1995
Rakennustaito 5/95, Helsinki 1995
Form Function 2/95, Helsinki 1995
Interni 12/95, Milan 1995
Arkkitehtuurikilpailuja 3/95, Helsinki 1995
Bau Magazin, Helsinki 1995
Teräsrakenne 3/95, Helsinki 1995
Projektiuutiset 2/95, Espoo 1995
Arkkitehtuurikilpailuja 2/95, Helsinki 1995
Le Carré Bleu 1/95, Paris 1995

Tango Mäntyniemi, Tasavallan Presidentin Virka-asunnon Arkkitehtuuri; Roger Connah, editor, Helsinki 1994
Pientalojen parhaat, Puuinfo Oy, Helsinki 1994
Rakennusteollisuus 3/94, Helsinki 1994
Lasirakentaja 1/94, Tampere 1994
Puu 1/94, Helsinki 1994
Teräsrakenne 1/94, Helsinki 1994
ATL, Arkkitehtitoimistot 1994-95, Helsinki 1994
Arkitekturtävlingar 10/93, Stockholm 1993
Arkkitehti 7-8/93, Helsinki 1993
Design in Finland, Helsinki 1993
Rakennussuunnittelun ammattikurssi:Katainen, TKK, Tampere 1993
Arkkitehtuurikilpailuja 6/93,Helsinki 1993
Uuden kirkkoarkkitehtuurin ongelmat: Jukka-Pekka Airas, Helsinki 1992
Public Design'92, Espoo 1992
Suomalaisia liikennemerkkejä, Helsinki 1992
A+U 5/92, Tokyo 1992
Suomi rakentaa 8, Helsinki 1992
KS Neues 1992, Munich 1992
Arkkitehti 4-5/92, Helsinki 1992
Innovations in Steel, Brussells 1992
Arkkitehtuurikilpailut, SAFA, Espoo 1992
Buhnentechnische Rundschau, June 1991
Arkkitehtuurikilpailuja 12/91, Helsinki 1991
Mitä Missä Milloin, Helsinki 1991
Teräsrakentaminen, TTKK, Tampere 1991
Takstooli 3/91, Helsinki 1991
Puu 2/91, Helsinki 1991
European Masters 3, Atrium, Barcelona 1991
Space & Place 3, Helsinki 1991
Muodon kuvat 1960-1990, Helsinki 1991
Prague Quadriennale'91 1991
AIT 5/91, Berlin 1991
Archis 1/91, Amsterdam 1991
Rakennusteollisuus 6-7/90, Helsinki 1990
Vuosisadan kaunein ?,Osmo Mikkonen, Helsinki 1990
Arkkitehti 8/90, Helsinki 1990
Rakennustuotanto 14.5.90, Helsinki 1990
Lighting+Sound, December 90 1990
Projektiuutiset 2/90, Helsinki 1990
Look at Finland 2/90, Helsinki 1990
L'Architecture d'aujourd'hui, avril 90, 268, Paris 1990
Teräsrakenne 2/90, 1990
Tiili 3/90, Helsinki 1990
Takstooli 1/90, Helsinki 1990
Rakennustaito 8/89, Helsinki 1989
Yliopiston Helsinki, Helsinki 1989
Arkkitehtuuri Itä-Suomessa, Rakentajain kustannus Oy, Jyväskylä 1989
Space Design 1/89, Tokyo 1989
Rakennustaito 4/88, Helsinki 1988
Design Management, Helsinki 1988
Building Design 891, April 1988
Teräsrakenne 4/88, Helsinki 1988
Imago ja identiteetti, Finnovatio 88, Lahti 1988
International Construction 2/88, Wallington 1988
Contact 4/87, Helsinki 1987
Tiili 4/87, Helsinki 1987
Nordic Sound, Sept. 87 1987
Julkisivu 3/87, Helsinki 1987
Ars Sacra Fennica, Helsinki 1987
Teräsrakenne 2/87, Helsinki 1987
Rakennushallitus 1811-1986, Helsinki 1986
Tiili 4/86, Helsinki 1986
Rakennushallinto 1961-1986, Helsinki 1986
Suomalainen pientalo, Helsinki 1986
Houttechniek 2/85, Amhem 1985
Funktion&Form: Charles von Büren, Basel 1985
Teräsrakenne 3/85. Helsinki 1985
KS Neues 2/84, Munich 1984
Kahri&Pyykönen: Asuntoarkkitehtuuri ja -suunnittelu, Rakennuskirja Oy, Helsinki 1984
Arkkitehtuurikilpailuja 4/84, Helsinki 1984
La Città Informa, Padova 1983
Architecture Minnesota 7-8/83, Minneapolis 1983
Idea 181/83, Tokyo 1983
Puu 2/83, Helsinki 1983
Progressive Architecture 9/83, Stamford 1983
Teräsrakenne 2/83, Helsinki 1983
A+U 10/83, Tokyo 1983
Arkkitehtuurikilpailuja 7/83, Helsinki 1983
Inland Architect 9-10/83 1983
Arkkitehtuurikilpailuja 7/82, Helsinki 1982
Arkkitehtuurikilpailuja 1/82, Helsinki 1982
Helsingin metro, Helsinki 1982
Esineitä ympärillämme, Tammi, Helsinki 1982
Arkkitehtuurikilpailuja 6-7/81, Helsinki 1981
Vuoden Huiput, Helsinki 1981
Arkkitehtuurikilpailuja 7/80, Helsinki 1980
Suomalaisia kytkettyjä pientaloia, TKK, Espoo 1980
Arkkitehtuurikilpailuja 7/79, Helsinki 1979
Arkkitehtuurikilpailuja 5-6/79, Helsinki 1979
Projekt 4/79, Warsaw 1979
Arkkitehtuurikilpailuja 4/79, Helsinki 1979
Domus 592/79, Milan 1979
Arkkitehtuurikilpailuja 1/79, Helsinki 1979
Space Design 8/78, Tokyo 1978
Novum Gebrauchsgraphik 7/78, Munich 1978
G4, imago, Helsinki 1977
Suomalaisia yhdenperheentaloja, TKK, Espoo 1977
Arkkitehti 3/76, Helsinki 1976
Suomalaisia loma-asuntoja ja saunoja, TKK, Espoo 1976
Arkkitehtuurikilpailuja 5-6/75, Helsinki 1975
Tiili 4/72, Helsinki 1972
Metropolis, NCSU, Raleigh 1972
Arkkitehtuurikilpailuja 3-4/71, Helsinki 1971
A+U 4/71, Tokyo 1971
Arkkitehti 8/70, Helsinki 1970
Suomi rakentaa 1965-1970, Helsinki 1970
Uima-altaat, W+G, Helsinki 1970
Arkkitehti 7/69, Helsinki 1969
Arkkitehtuurikilpailuja 3-4/69, Helsinki 1969
Avotakka 4/69, Helsinki 1969
Graphis Annual 1968-69, Zürich 1969
Arkkitehti 7/68, Helsinki 1968
Graphis Annual 1967-68, Zürich 1968
Taidetapahtuma 1, Turku 1966

OATH OF OFFICE 1973

LANGUAGES
Certificate of English, TOEFL 1971
Cerificate of Swedish, Statens språkkkexamensnämnd 2002

MILITARY SERVICE

Second lieutenant 1963

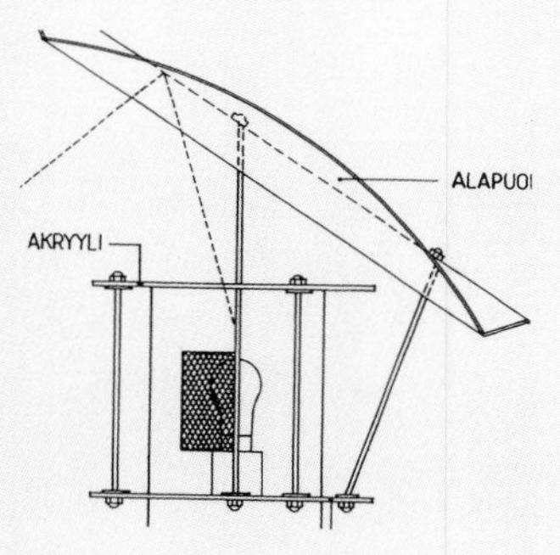

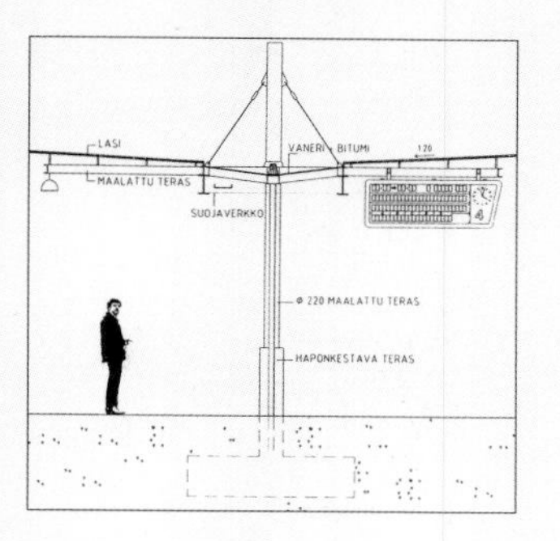

Design teams

设计队伍

Arkkitehtitoimisto Esa Piironen and Mikko Pulkkinen
1966～1971

Eeva Paasimaa
Esa Piironen
Mikko Pulkkinen
Frank Schauman
Jyrki Tasa

Suunnittelutoimisto G4
1970-85

Martin Bovill
Pirkko Eskola
Tuure Haikonen
Leena Hakotie
Marjatta Helenius
Martti Huttunen
Anitta Kainulainen
Marja Koponen
Antti Laiho
Ola Laiho
Inkeri Laine
Rainer Lindberg
Juha Lumme
Lasse Lyytikäinen
Juhani Maunula
Esko Miettinen
Jyrki Nieminen
Timo Paasimaa
Juhani Pallasmaa
Esa Piironen
Pirkko Rehfeld
Vesa Rosilo
Tuija Siivonen
Maija Tomevski
Vladimir Tomevski
Juhani Väisänen

Arkkitehtitoimisto Sakari Aartelo and Esa Piironen
1978～1990

Sakari Aartelo
Erkki Forsman
Mikko Helasvuo
Pertti Hokkanen
Hannu Huttunen
Katri Jaatinen
Jutta Johansson
Olli Karttunen
Lasse Kojo
Antti Korkkula
Antti Laiho
Inkeri Laine
Henrik Lares
Eero Lehto
Olavi Lipponen
Juha Lumme
Leena Makkonen
Nina Mattila
Peter Meyer
Harri Ojala
Timo Paasimaa
Mika Penttinen
Annukka Piironen
Esa Piironen
Juha Poutanen
Reijo Pulkkinen
Tiina Riihimäki
Pirkko Varila
Mika Westerback
Kirsi Vuori
Leena Yli-Lonttinen
Anu Ylitalo

Arkkitehtitoimisto Esa Piironen Oy
1990

Harri Andersson
Inkeri Brigatti
Teemu Halme
Atte Hilanka
Vesa Jäntti
Mikko Kalkkinen
Tommi Kantonen
Jari Lepistö
Mari Mannevaara
Ulla Manninen
Peter Meyer
Harri Ojala
Timo Paasimaa
Erkki Piironen
Esa Piironen
Ulla-Maija Raappana
Arja Rahiala
Roope Rissanen
Kaarlo Rohola
Petri Saarelainen
Kai Salmi
Sami Vikström
Jarmo Viljakka
Matti Vänskä

Photographers

Kari Hakli
Ola Laiho
Voitto Niemelä
Aulis Nyqvist
Esa Piironen
Simo Rista
Seppo Saves
Jussi Tiainen
Harri Toivola

Model makers

Martti Helispää
Aukusti Huoponen
Jari Jetsonen
Olli-Pekka Keramaa
Niilo Vainio